Geography of Innovation

Within the European context of innovation for growth, public and corporate actors are faced with pressing questions concerning innovation policy and the return on public and private investment in innovation at the regional level. To help them answer these questions, researchers in the field of Geography of Innovation propose interesting developments and new perspectives for the analysis of localized innovation processes, interactions between science, technology and industry, and their impact on regional growth and competitiveness, offering new foundations for designing and evaluating public policies.

The aim of this book is firstly to highlight major recent methodological advances in the Geography of Innovation, particularly concerning the measurement of spatial knowledge externalities and their impact on agglomeration effects. Strategic approaches using microeconomic data have also contributed to showing how firms' strategies may interact with the local environment and impact upon agglomeration dynamics.

Interesting new results emerge from the application of these new methodologies to the analysis of innovation dynamics in European regions and this book shows how they can help revisit some of the main tenets of received wisdom concerning the rationale and impact of public policies on the Geography of Innovation. This book was previously published as a special issue of *Regional Studies*.

Nadine Massard is Professor in economics at the University Grenoble Alpes (France) and member of Grenoble Applied Economics Laboratory (GAEL, INRA, CNRS). Her main research theme is Economic Geography of Innovation and Public Policy. Currently she is the President of EuroLIO (European Localized Innovation Observatory).

Corinne Autant-Bernard is Professor in economics at Saint-Etienne University. Author of dozens of articles and book chapters in economics of innovation, she studies spatial concentration and networks of innovation and their impacts on productivity. She also contributes in technological transfer, providing policy makers with localized innovation diagnostics and policy evaluation.

Regions and Cities

Series Editor in Chief
Susan M. Christopherson, *Cornell University, USA*

Editors
Maryann Feldman, *University of Georgia, USA*
Gernot Grabher, *HafenCity University Hamburg, Germany*
Ron Martin, *University of Cambridge, UK*
Kieran P. Donaghy, *Cornell University, USA*

In today's globalised, knowledge-driven and networked world, regions and cities have assumed heightened significance as the interconnected nodes of economic, social and cultural production, and as sites of new modes of economic and territorial governance and policy experimentation. This book series brings together incisive and critically engaged international and interdisciplinary research on this resurgence of regions and cities, and should be of interest to geographers, economists, sociologists, political scientists and cultural scholars, as well as to policy-makers involved in regional and urban development.

For more information on the Regional Studies Association visit www.regionalstudies.org

112. The Empirical and Institutional Dimensions of Smart Specialisation
Edited by John Goddard, Philip McCann and Frank Van Oort

111. EU Cohesion Policy
Reassessing performance and direction
Edited by John Bachtler, Peter Berkowitz, Sally Hardy and Tatjana Muravska

110. Geography of Innovation
Edited by Nadine Massard and Corinne Autant-Bernard

109. Rethinking International Skilled Migration
Edited by Micheline van Riemsdijk and Qingfang Wang

108. The EU's New Borderland
Cross-border relations and regional development
Andrzej Jakubowski, Andrzej Miszczuk, Bogdan Kawałko, Tomasz Komornicki, and Roman Szul

107. Entrepreneurship in a Regional Context
Edited by Michael Fritsch and David J. Storey

106. Governing Smart Specialisation
Edited by Dimitrios Kyriakou, Manuel Palazuelos Martínez, Inmaculada Periáñez-Forte, and Alessandro Rainoldi

105. Innovation, Regional Development and the Life Sciences
Beyond clusters
Kean Birch

104. Unfolding Cluster Evolution
Edited by Fiorenza Belussi and Jose Luis Hervás-Olivier

103. Place-based Economic Development and the New EU Cohesion Policy
Edited by Philip McCann and Attila Varga

102. Transformation of Resource Towns and Peripheries
Political economy perspectives
Edited by Greg Halseth

101. Approaches to Economic Geography
Towards a geographical political economy
Ray Hudson

100. Secondary Cities and Development
Edited by Lochner Marais, Etienne Nel and Ronnie Donaldson

99. Technology and the City
Systems, applications and implications
Tan Yigitcanlar

98. Smaller Cities in a World of Competitiveness
Peter Karl Kresl and Daniele Ietri

97. Code and the City
Edited by Rob Kitchin and Sung-Yueh Perng

96. The UK Regional–National Economic Problem
Geography, globalisation and governance
Philip McCann

95. Skills and Cities
Edited by Sako Musterd, Marco Bontje and Jan Rouwendal

94. Higher Education and the Creative Economy
Beyond the campus
Edited by Roberta Comunian and Abigail Gilmore

93. Making Cultural Cities in Asia
Mobility, assemblage, and the politics of aspirational urbanism
Edited by Jun Wang, Tim Oakes and Yang Yang

92. Leadership and the City
Power, strategy and networks in the making of knowledge cities
Markku Sotarauta

91. Evolutionary Economic Geography
Theoretical and empirical progress
Edited by Dieter Kogler

90. Cities in Crisis
Socio-spatial impacts of the economic crisis in Southern European cities
Edited by Jörg Knieling and Frank Othengrafen

89. Socio-Economic Segregation in European Capital Cities
East meets West
Edited by Tiit Tammaru, Szymon Marcińczak, Maarten van Ham, Sako Musterd

88. People, Places and Policy
Knowing contemporary Wales through new localities
Edited by Martin Jones, Scott Orford and Victoria Macfarlane

87. The London Olympics and Urban Development
The mega-event city
Edited by Gavin Poynter, Valerie Viehoff and Yang Li

86. Making 21st Century Knowledge Complexes
Technopoles of the world revisited
Edited by Julie Tian Miao, Paul Benneworth and Nicholas A. Phelps

85. Soft Spaces in Europe
Re-negotiating governance, boundaries and borders
Edited by Philip Allmendinger, Graham Haughton, Jörg Knieling and Frank Othengrafen

84. Regional Worlds: Advancing the Geography of Regions
Edited by Martin Jones and Anssi Paasi

83. Place-making and Urban Development
New challenges for contemporary planning and design
Pier Carlo Palermo and Davide Ponzini

82. Knowledge, Networks and Policy
Regional studies in postwar Britain and beyond
James Hopkins

81. Dynamics of Economic Spaces in the Global Knowledge-based Economy
Theory and East Asian cases
Sam Ock Park

80. Urban Competitiveness
Theory and practice
Daniele Letri and Peter Kresl

79. Smart Specialisation
Opportunities and challenges for regional innovation policy
Dominique Foray

78. The Age of Intelligent Cities
Smart environments and innovation-for-all strategies
Nicos Komninos

77. Space and Place in Central and Eastern Europe
Historical trends and perspectives
Gyula Horváth

76. Territorial Cohesion in Rural Europe
The relational turn in rural development
Edited by Andrew Copus and Philomena de Lima

75. The Global Competitiveness of Regions
Robert Huggins, Hiro Izushi, Daniel Prokop and Piers Thompson

74. The Social Dynamics of Innovation Networks
Edited by Roel Rutten, Paul Benneworth, Dessy Irawati and Frans Boekema

73. The European Territory
From historical roots to global challenges
Jacques Robert

72. Urban Innovation Systems
What makes them tick?
Willem van Winden, Erik Braun, Alexander Otgaar and Jan-Jelle Witte

71. Shrinking Cities
A global perspective
Edited by Harry W. Richardson and Chang Woon Nam

70. Cities, State and Globalization
City-regional governance
Tassilo Herrschel

69. The Creative Class Goes Global
Edited by Charlotta Mellander, Richard Florida, Bjørn Asheim and Meric Gertler

68. Entrepreneurial Knowledge, Technology and the Transformation of Regions
Edited by Charlie Karlsson, Börje Johansson and Roger Stough

67. The Economic Geography of the IT Industry in the Asia Pacific Region
Edited by Philip Cooke, Glen Searle and Kevin O'Connor

66. Working Regions
Reconnecting innovation and production in the knowledge economy
Jennifer Clark

65. Europe's Changing Geography
The impact of inter-regional networks
Edited by Nicola Bellini and Ulrich Hilpert

64. The Value of Arts and Culture for Regional Development
A Scandinavian perspective
Edited by Lisbeth Lindeborg and Lars Lindkvist

63. The University and the City
John Goddard and Paul Vallance

62. Re-framing Regional Development
Evolution, innovation and transition
Edited by Philip Cooke

61. Networking Regionalised Innovative Labour Markets
Edited by Ulrich Hilpert and Helen Lawton Smith

60. Leadership and Change in Sustainable Regional Development
Edited by Markku Sotarauta, Ina Horlings and Joyce Liddle

59. Regional Development Agencies: The Next Generation?
Networking, knowledge and regional policies
Edited by Nicola Bellini, Mike Danson and Henrik Halkier

58. Community-based Entrepreneurship and Rural Development
Creating favourable conditions for small businesses in Central Europe
Matthias Fink, Stephan Loidl and Richard Lang

57. Creative Industries and Innovation in Europe
Concepts, measures and comparative case studies
Edited by Luciana Lazzeretti

56. Innovation Governance in an Open Economy
Shaping regional nodes in a globalized world
Edited by Annika Rickne, Staffan Laestadius and Henry Etzkowitz

55. Complex Adaptive Innovation Systems
Relatedness and transversality in the evolving region
Philip Cooke

54. Creating Knowledge Locations in Cities
Innovation and integration challenges
Willem van Winden, Luis de Carvalho, Erwin van Tujil, Jeroen van Haaren and Leo van den Berg

53. Regional Development in Northern Europe
Peripherality, marginality and border issues
Edited by Mike Danson and Peter De Souza

52. Promoting Silicon Valleys in Latin America
Luciano Ciravegna

51. Industrial Policy Beyond the Crisis
Regional, national and international perspectives
Edited by David Bailey, Helena Lenihan and Josep-Maria Arauzo-Carod

50. Just Growth
Inclusion and prosperity in America's metropolitan regions
Chris Benner and Manuel Pastor

49. Cultural Political Economy of Small Cities
Edited by Anne Lorentzen and Bas van Heur

48. The Recession and Beyond
Local and regional responses to the downturn
Edited by David Bailey and Caroline Chapain

47. Beyond Territory
Edited by Harald Bathelt, Maryann Feldman and Dieter F. Kogler

46. Leadership and Place
Edited by Chris Collinge, John Gibney and Chris Mabey

45. Migration in the 21st Century
Rights, outcomes, and policy
Kim Korinek and Thomas Maloney

44. The Futures of the City Region
Edited by Michael Neuman and Angela Hull

43. The Impacts of Automotive Plant Closures
A tale of two cities
Edited by Andrew Beer and Holli Evans

42. Manufacturing in the New Urban Economy
Willem van Winden, Leo van den Berg, Luis de Carvalho and Erwin van Tuijl

41. Globalizing Regional Development in East Asia
Production networks, clusters, and entrepreneurship
Edited by Henry Wai-chung Yeung

40. China and Europe
The implications of the rise of China as a global economic power for Europe
Edited by Klaus Kunzmann, Willy A Schmid and Martina Koll-Schretzenmayr

39. Business Networks in Clusters and Industrial Districts
The governance of the global value chain
Edited by Fiorenza Belussi and Alessia Sammarra

38. Whither Regional Studies?
Edited by Andy Pike

37. Intelligent Cities and Globalisation of Innovation Networks
Nicos Komninos

36. Devolution, Regionalism and Regional Development
The UK experience
Edited by Jonathan Bradbury

35. Creative Regions
Technology, culture and knowledge entrepreneurship
Edited by Philip Cooke and Dafna Schwartz

34. European Cohesion Policy
Willem Molle

33. Geographies of the New Economy
Critical reflections
Edited by Peter W. Daniels, Andrew Leyshon, Michael J. Bradshaw and Jonathan Beaverstock

32. The Rise of the English Regions?
Edited by Irene Hardill, Paul Benneworth, Mark Baker and Leslie Budd

31. Regional Development in the Knowledge Economy
Edited by Philip Cooke and Andrea Piccaluga

30. Regional Competitiveness
Edited by Ron Martin, Michael Kitson and Peter Tyler

29. Clusters and Regional Development
Critical reflections and explorations
Edited by Bjørn Asheim, Philip Cooke and Ron Martin

28. Regions, Spatial Strategies and Sustainable Development
David Counsell and Graham Haughton

27. Sustainable Cities
Graham Haughton and Colin Hunter

26. Geographies of Labour Market Inequality
Edited by Ron Martin and Philip S. Morrison

25. Regional Innovation Strategies
The challenge for less-favoured regions
Edited by Kevin Morgan and Claire Nauwelaers

24. Out of the Ashes?
The social impact of industrial contraction and regeneration on Britain's mining communities
Chas Critcher, Bella Dicks, David Parry and David Waddington

23. Restructuring Industry and Territory
The experience of Europe's regions
Edited by Anna Giunta, Arnoud Lagendijk and Andy Pike

22. Foreign Direct Investment and the Global Economy
Corporate and institutional dynamics of global-localisation
Edited by Jeremy Alden and Nicholas F. Phelps

21. Community Economic Development
Edited by Graham Haughton

20. Regional Development Agencies in Europe
Edited by Charlotte Damborg, Mike Danson and Henrik Halkier

19. Social Exclusion in European Cities
Processes, experiences and responses
Edited by Judith Allen, Goran Cars and Ali Madanipour

18. Metropolitan Planning in Britain
A comparative study
Edited by Peter Roberts, Kevin Thomas and Gwyndaf Williams

17. Unemployment and Social Exclusion
Landscapes of labour inequality and social exclusion
Edited by Sally Hardy, Paul Lawless and Ron Martin

16. Multinationals and European Integration
Trade, investment and regional development
Edited by Nicholas A. Phelps

15. The Coherence of EU Regional Policy
Contrasting perspectives on the structural funds
Edited by John Bachtler and Ivan Turok

14. New Institutional Spaces
TECs and the remaking of economic governance
Martin Jones, Foreword by Jamie Peck

13. Regional Policy in Europe
S. S. Artobolevskiy

12. Innovation Networks and Learning Regions?
James Simmie

11. British Regionalism and Devolution
The challenges of state reform and European integration
Edited by Jonathan Bradbury and John Mawson

10. Regional Development Strategies
A European perspective
Edited by Jeremy Alden and Philip Boland

9. Union Retreat and the Regions
The shrinking landscape of organised labour
Ron Martin, Peter Sunley and Jane Wills

8. The Regional Dimension of Transformation in Central Europe
Grzegorz Gorzelak

7. The Determinants of Small Firm Growth
An inter-regional study in the United Kingdom 1986–90
Richard Barkham, Graham Gudgin, Mark Hart and Eric Hanvey

6. The Regional Imperative
Regional planning and governance in Britain, Europe and the United States
Urlan A. Wannop

5. An Enlarged Europe
Regions in competition?
Edited by Louis Albrechts, Sally Hardy, Mark Hart and Anastasios Katos

4. Spatial Policy in a Divided Nation
Edited by Richard T. Harrison and Mark Hart

3. Regional Development in the 1990s
The British Isles in transition
Edited by Ron Martin and Peter Townroe

2. Retreat from the Regions
Corporate change and the closure of factories
Stephen Fothergill and Nigel Guy

1. Beyond Green Belts
Managing urban growth in the 21st century
Edited by John Herington

Geography of Innovation

New Trends and Implication for Public Policy Renewal

Edited by
Nadine Massard and Corinne Autant-Bernard

Routledge
Taylor & Francis Group
LONDON AND NEW YORK

First published 2018
by Routledge
2 Park Square, Milton Park, Abingdon, Oxon, OX14 4RN, UK

and by Routledge
711 Third Avenue, New York, NY 10017, USA

Routledge is an imprint of the Taylor & Francis Group, an informa business

British Library Cataloguing in Publication Data
A catalogue record for this book is available from the British Library

ISBN 13: 978-1-138-20914-5

Typeset in Bembo
by RefineCatch Limited, Bungay, Suffolk

Publisher's Note
The publisher accepts responsibility for any inconsistencies that may have arisen during the conversion of this book from journal articles to book chapters, namely the possible inclusion of journal terminology.

Contents

Citation Information

The chapters in this book were originally published in *Regional Studies*, volume 49, issue 11 (November 2015). When citing this material, please use the original page numbering for each article, as follows:

Chapter 1
Geography of Innovation: New Trends and Implications for Public Policy Renewal
Nadine Massard and Corinne Autant-Bernard
Regional Studies, volume 49, issue 11 (November 2015), pp. 1767–1771

Chapter 2
Regional Heterogeneity and Interregional Research Spillovers in European Innovation: Modelling and Policy Implications
Gianni Guastella and Frank G. Van Oort
Regional Studies, volume 49, issue 11 (November 2015), pp. 1772–1787

Chapter 3
Knowledge, Innovation and Productivity Gains across European Regions
Roberta Capello and Camilla Lenzi
Regional Studies, volume 49, issue 11 (November 2015), pp. 1788–1804

Chapter 4
Do Technology Leaders Deter Inward R&D Investments? Evidence from Regional R&D Location Decisions in Europe
René Belderbos and Dieter Somers
Regional Studies, volume 49, issue 11 (November 2015), pp. 1805–1821

Chapter 5
Industry-Specific Firm Growth and Agglomeration
Matthias Duschl, Tobias Scholl, Thomas Brenner, Dennis Luxen and Falk Raschke
Regional Studies, volume 49, issue 11 (November 2015), pp. 1822–1839

Chapter 6
Marshall's versus Jacobs' Externalities in Firm Innovation Performance: The Case of French Industry
Danielle Galliano, Marie-Benoît Magrini and Pierre Triboulet
Regional Studies, volume 49, issue 11 (November 2015), pp. 1840–1858

Chapter 7
Regional Knowledge Flows and Innovation Policy: A Dynamic Representation
Ugo Fratesi
Regional Studies, volume 49, issue 11 (November 2015), pp. 1859–1872

For any permission-related enquiries please visit:
http://www.tandfonline.com/page/help/permissions

Notes on Contributors

René Belderbos is Professor in the Department of Managerial Economics, Strategy and Innovation, Faculty of Economics and Business, University of Leuven, Belgium.

Corinne Autant-Bernard is Professor in economics at Saint-Etienne University, France.

Thomas Brenner is Professor in the Department of Geography, Philipps University of Marburg, Germany.

Roberta Capello is Professor of Regional Economics at Politecnico di Milano, Milan, Italy.

Matthias Duschl is Post-Doctoral Research Fellow in the Department of Geography, Philipps University of Marburg, Germany.

Ugo Fratesi is Associate Professor in the Department of Architecture, Built environment and Construction Engineering, Politecnico di Milano, Milan, Italy.

Danielle Galliano is Associate Director at INRA, Castanet Tolosan, France.

Gianni Guastella is Post-Doctoral Research Fellow in the Department of Mathematics and Physics, Università Cattolica, Brescia, Italy.

Camilla Lenzi is Associate Professor in the Department of Architecture, Built environment and Construction Engineering, Politecnico di Milano, Milan, Italy.

Dennis Luxen is a researcher in the Department of Informatics, Institute of Theoretical Informatics, Karlsruhe Institute of Technology, Germany.

Marie-Benoît Magrini is an economist at INRA, Castanet Tolosan, France.

Nadine Massard is Professor in economics at the University Grenoble Alpes, France, and member of Grenoble Applied Economics Laboratory (GAEL, INRA, CNRS).

Falk Raschke is a researcher in the Department of Geography, Philipps University of Marburg, Frankfurt, Germany.

Tobias Scholl is a researcher in the Department of Geography, Philipps University of Marburg, Frankfurt, Germany.

Dieter Somers is a researcher in the Department of Managerial Economics, Strategy and Innovation, Faculty of Economics and Business, University of Leuven, Belgium.

Pierre Triboulet is a research engineer at INRA, Castanet Tolosan, France.

Frank G. Van Oort is Professor of Urban & Regional Economics in the Erasmus School of Economics Institute for Housing and Urban Development Studies (IHS), Erasmus University, Rotterdam, the Netherlands.

Geography of Innovation: New Trends and Implications for Public Policy Renewal

NADINE MASSARD†‡ and CORINNE AUTANT-BERNARD§# ¶
†INRA, GAEL, Grenoble, France.
‡University of Grenoble Alpes, GAEL, Grenoble, France
§University of Lyon, Lyon, France
#CNRS, GATE Lyon Saint-Etienne, Ecully, France
¶Jean Monnet University, Saint-Etienne, France

THE GEOGRAPHY OF INNOVATION AND PUBLIC POLICIES: FROM PLACE-NEUTRAL TO PLACE-BASED POLICIES

Within the European context of innovation for growth, public and corporate actors are faced with pressing questions concerning innovation policy and the return on public and private investment in innovation, particularly at the local/regional level. To help them answer these questions, the two first European conferences on the 'Geography of Innovation' in St-Etienne (France, 2012) and Utrecht (the Netherlands, 2014) highlighted interesting developments and new perspectives for the analysis of localized innovation processes, interactions between science, technology and industry, and their impact on regional growth and competitiveness, offering new bases for designing and evaluating public policies.

Estimating a territorial knowledge production function (KPF) using aggregated data at the regional level is one of the main analytical tools traditionally used in the Geography of Innovation field. Starting from the seminal volume published by Maryann Feldman in the mid-1990s (FELDMAN, 1994), the numerous studies that have used such empirical tools have progressively and significantly improved measurement of the geographical dimension of knowledge externalities, introducing more complex processes of knowledge diffusion and using new spatial econometric tools.

Although such improvements have contributed to refining the modelling of spatial externalities (ACS *et al.*, 2002; AUTANT-BERNARD, 2012; FINGLETON and LOPEZ-BAZO, 2006), these studies nevertheless still fail to offer a reliable picture of the diverse dimensions of localized innovation processes and dynamics. In particular, although it is possible to draw interesting general implications for local public policies from this literature (AUTANT-BERNARD *et al.*, 2014), key issues remain that are known to be potential sources of ambiguous or contradictory results concerning the effectiveness of policies implemented in regions with very diverse profiles. Hence, literature reviews in this field often agree on the need to go beyond general conclusions in terms of agglomeration economies and to develop specific policies taking into account the very particular characteristics of the local context (AUTANT-BERNARD *et al.*, 2014).

Indeed, while the heterogeneity of industry and the importance of the chosen regional level (NUTS-1, -2 or -3 level, city or cluster-level data) have already been emphasized to explain some differences in the empirical results obtained, it is worth noting that, to date, the capacity of studies that use spatially extended KPFs to address the question of the heterogeneity of the concerned territories remain limited.

Yet, this question is central to helping policy-makers switch from simple place-neutral policies to designing place-based policies (BARCA *et al.*, 2012), for example in the sense defined by the European debate on the regional 'smart specialization' strategy (EUROPEAN COMMISSION, 2012; MCCANN and ORTEGA-ARGILES, 2015). New methodologies should be developed not only to identify regularities in spatial knowledge diffusion and production mechanisms, but also, beyond this, to develop these regularities and qualify them in a world characterized by the existence of diverse regional contexts for innovation.

Another main limit of using a territorial KPF perspective as the foundation for place-based policies lies in its inability to take into account the importance of firms' behaviours and their relation to location. Public actors, especially at the regional level, not only have to deal with public research and development (R&D) infrastructures and expenditure, but also they have to consider the attractiveness of their regions for private R&D investment and the creation/exploitation of innovation. Providing consistency between regional structure and dynamics with private firms' strategies is no easy task. Nor is it straightforward to design the best incentives for the development of private innovation activities. A better understanding of the interaction between the local environment and firms' dynamics is required and the main attempts to do so rely on the application of micro-econometric techniques to individual data.

Hence, efforts to shift from aggregated large-scale investigations to micro-economic data-driven approaches are needed to qualify better the common belief in the connection between agglomeration, externalities and positive impact on innovation, productivity and growth. Possible negative effects due to competitive pressure or to firms' internal strategies have to be thoroughly investigated to help policymakers design more efficient incentives.

Considering these new developments in analytical tools and the improvement of micro-geolocalized databases which become ever more international and inter-temporal, it is now possible for Geography of Innovation researchers not only to provide general rationales for public policies accounting for the regional dimension of knowledge flows and agglomeration economies, but also to help define and design local solutions within a global world where spatial interdependencies act on different levels. Indeed public policies based on the Geography of Innovation are at a crossroads: either they only emphasize 'agglomeration economies' resulting from the co-location of innovative actors calling for ever-growing urban agglomerations, or they effectively consider the great diversity of regional contexts and the diversity in spatial organization for innovation, hence supporting the idea that polarization in very large urban agglomerations is not the only model of spatial organization likely to favour innovation production and diffusion. In this case, however, much remains to be done to help each region be innovative in its policy design and to find the policy mix that best corresponds to its own strategic needs and position in a global economy.

OBJECTIVES OF THIS SPECIAL ISSUE

The aim of this special issue is threefold. Firstly, it highlights major recent methodological advances to address the two key issues referred to above: improving extended KPF analyses, on the one hand, and developing strategic approaches using microeconomic data, on the other:

- Two papers are presented using KPFs. They offer new methodologies to deal with the issue of regional heterogeneity when estimating KPF at the regional level in Europe. Although it is acknowledged that economic agents within each region may have a diversity of innovative behaviours, these first papers aim to guide innovation strategies at the regional level. Their value thus lies in obtaining results on average innovative behaviour at the regional level, a spatial unit particularly concerned by smart specialization strategies.
- Using more microeconomic approaches, three papers contribute to the second topic. They use micro-economic data to show how firms' strategies may interact with the local environment and impact upon the determinants of agglomeration dynamics.

The special issue concludes with a paper presenting a new theoretical framework for the analysis and evaluation of local innovation public policies using simulation methodology. All these papers have important policy implications, discussed below.

Secondly, this issue draws attention to interesting new results emerging from the application of these new methodologies to the analysis of innovation dynamics in European regions and shows how they can help one to revisit some main tenets of received wisdom concerning the rationale and impact of public policies on the Geography of Innovation.

Finally this special issue also identifies issues that still require further research, particularly in relation to the development of new methodologies for the evaluation of public policies integrating the spatial dimension and the interdependencies between public policies implemented at different regional scales, which remains no more than an emerging field in the Geography of Innovation.

CONTENT OF THIS SPECIAL ISSUE

To analyse these questions, this special issue presents five empirical articles using econometric techniques and one theoretical article using simulation techniques.

GUASTELLA and VAN OORT (2015, in this issue) provide a good introduction to this special issue with a fairly comprehensive review of the literature on KPF and the criticism and debates which empirical work in this field since the 1990s has encountered, particularly in terms of its ability to account for spatial heterogeneity. Using a spatially extended KPF, this paper examines the contribution of research spillovers to regional innovation in the European Union. The question of spatial heterogeneity is addressed through the use of a generalized additive model in which a non-linear spatial trend is coupled with the standard patent equation. The main finding is that once spatial heterogeneity is accounted for, evidence of spillovers substantially decreases. Hence, omitting region-specific characteristics in the econometric specification introduces bias into the R&D-related coefficient, which has often led to overestimating spillover effects in earlier literature.

The paper by CAPELLO and LENZI (2014, in this issue) goes further towards modelling spatial heterogeneity on innovation processes. The way in which knowledge and innovation merge and are supported and stimulated by specific territorial preconditions generates what the authors call a 'territorial pattern of innovation'. The conceptual expectation is that while productivity accruing from knowledge is highly selective in space and concentrated in a handful of regions already showing greater knowledge endowment, the benefits of innovation are more spatially dispersed. For this reason, the distinctions between scientific versus non-scientific knowledge, on the one hand, and knowledge versus innovation, on the other, drive the analysis. The relationship between knowledge (i.e. patent intensity), innovation and economic efficiency (i.e. total factor productivity – TFP) is studied at the European Union regional level (NUTS-2). The results confirm that while both knowledge and innovation are important drivers of economic efficiency, only regions strongly endowed with elements supporting knowledge creation processes are likely to benefit from the positive returns to knowledge. Moreover, while the benefits resulting from knowledge appear to be rather selective and concentrated in a relatively small number of regions, the benefits accruing from innovation seem more pervasive and beneficial to a larger number of regions.

The question posed by BELDERBOS and SOMERS (2015, in this issue) is of crucial interest for local policy-makers. Drawing on data provided by the *Financial Times*' Cross-border Investment Monitor (2003–08) on 196 foreign R&D investment projects in EU-15 countries at the NUTS-1 regional level, this paper suggests that while the strength and quality of relevant local technology development activities attract cross-border R&D investments, investors are discouraged by concentrated technology development activities due to the dominance of regional industry leaders. Indeed, asymmetry in knowledge spillovers between technology leading and lagging firms provides incentives for technology leaders to discourage other firms from setting up new R&D establishments in their vicinity in order to reduce the threat of knowledge dissipation. This discouraging effect occurs specifically when leaders exhibit internally oriented R&D organization, drawing on cross-border intra-firm knowledge flows in their technology development activities. Such results challenge the common premise in the literature on clusters that firms will generally benefit from knowledge spillovers when locating in regions with clustered industrial activity, while highlighting that geographical clustering and knowledge spillovers do not necessarily co-occur, but that knowledge spillovers are conditional on incumbent firms' strategic behaviour.

DUSCHL *et al.* (2014, in this issue) address the impact of clusters on individual firm growth rates with a more methodological perspective, the objective of which is to refine the geographical analysis of cluster and better account for industrial specificities. The effects of agglomeration on employment and innovation are examined by geolocating firms into a more realistic relational space using travel time distances and flexible distance-decay function specifications. Results from quantile regression techniques on a large sample of German firms indicate that the spatial characteristics of these effects differ substantially between different types of industries and agglomeration effects. Being located in agglomerations of own-industry employment does not increase but rather reduces firms' growth prospects. In contrast, the impact of being located in proximity to knowledge-generating activities depends more systematically on the type and age of the industry – in less mature, more knowledge-intensive industries the relationship with firm growth tends to be positive. The spatial scale also varies across industries. In order to account for the heterogeneity of the analysed industries, results from three representative cases are also discussed in more depth and show that the very notion of cluster and its spatial dimension should be industry based.

In the article by GALLIANO *et al.* (2014, in this issue), the localization of firms is much more specific because it is based on a broader definition of the multi-establishment firm for which spatial externalities depend not only on the location of the headquarters but also on the location of other establishments. Hence, the authors test the influence of spatial externalities on firm innovation according to two ideas of the firm: considering its headquarters alone or considering the spatial environment of the majority of its employees. The Marshall/Jacobs dichotomy is renewed by taking into account the complexity of the different types of environment in which a firm could be located. Using a dataset on French industrial firms (Community Innovation Survey, CIS8) and different indicators to evaluate the specialization and/or diversification of all employment zones in France, the authors demonstrate that firms' geographic environments do not have the same effects on probability to innovate, on the one hand, and on innovation intensity, on the other. The effects also vary depending on the firm's organizational structure. The diversified area-based model tends to reveal the stimulating significance of 'Jacobian' diversity on firms' decisions to engage in innovation and its negative effect on their innovation intensity. In contrast, the specialized area-based model shows that being located close to firms in the same industry has little influence on the decision to innovate and reinforces innovation intensity. Finally, the dual model confirms these results by emphasizing the combined effects of specialization and diversification on innovation.

The last paper in this special issue is by FRATESI (2015, in this issue), who presents a first step towards the study of the effects of innovation policies on regional growth and income through its impacts on intraregional and interregional knowledge flows. The aim of the paper is to show that different types of innovation policy should be used in different regional contexts. A new tool for policy-makers to *ex ante* assess place-based policies is proposed. Drawing on recent theoretical advances in evolutionary economic geography (BOSCHMA and FRENKEN, 2006), an evolutionary economic geography simulation model is built that models knowledge flows within a region and between the region and external regions, based on the presence and mobility of the most relevant economic agents, i. e., basic and applied researchers, firms/entrepreneurs and qualified workers. The model is calibrated on an average NUTS-2 European region. Simulation methodology makes it possible to compare a number of possible policy interventions which could be applied to the region in terms of magnitude and time delay. By modelling different regional situations using different parameters of the model, the paper also shows that the same policy can have very different impacts in different contexts. The framework is general enough to be extended to analyse region- and sector-specific policies or to disaggregate regional institutions.

TO CONCLUDE

When considered together, the articles gathered in this special issue provide interesting results and insights concerning the current research and show the progress recently made in the Geography of Innovation field. Such progress now makes it possible to support a significant renaissance in the design of innovation policies, taking better account of local contexts and firms' strategies. However, the success of these policies in a European context requires real acknowledgement of the multi-scale dimension of interventions. The definition of local policies cannot be removed from the spatial aspect of national and European policies on innovation. Going beyond the issue of data, theoretical analysis frameworks are still lacking to shed light upon vertical (region, nation, Europe) or horizontal (between regions) interdependencies between public policies and multilevel impact diffusion phenomena.

Finally, another area merits further empirical research. The complexity of processes connecting diffusion of externalities, innovation, impact on firms' economic performance and that of the regions is increasingly informed. Nevertheless, it has to be acknowledged that the scope of Geography of Innovation has not yet succeeded in demonstrating solid results in terms of the temporality of these observed effects. Little is known for instance about the time lag between knowledge creation and its effects on innovation as well as on the speed of knowledge diffusion through space.

These multi-scale and temporal analytical questions pose major challenges for future research into the Geography of Innovation in that they determine one's ability to develop relevant frameworks for the *ex ante* and *ex post* evaluation of place-based innovation policies.

Disclosure statement – No potential conflict of interest was reported by the guest editors.

Funding – This work was supported by the Rhône-Alpes Regional Council [grant number ARC8].

REFERENCES

ACS Z., ANSELIN L. and VARGA A. (2002) Patents and innovation counts as measures of regional production of new knowledge, *Research Policy* **31**, 1069–1085. doi:10.1016/S0048-7333(01)00184-6

AUTANT-BERNARD C. (2012) Spatial econometrics of innovation: recent contributions and research perspectives, *Spatial Economic Analysis* **7**, 403–419. doi:10.1080/17421772.2012.722665

AUTANT-BERNARD C., MASSARD N. and COWAN R. (2014) Editors' Introduction to Spatial Knowledge Networks: Structure, driving forces and innovative performances, *Annals of Regional Science* **53**, 315–323. doi:10.1007/s00168-014-0642-0

BARCA F., MCCANN P. and RODRÍGUEZ-POSE A. (2012) The case for regional development intervention: place-based versus place-neutral approaches, *Journal of Regional Science* **52**, 134–152. doi:10.1111/j.1467-9787.2011.00756.x

BELDERBOS R. and SOMERS D. (2015) Do technology leaders deter inward R&D investments? Evidence from regional R&D location decisions in Europe, *Regional Studies*. doi:10.1080/00343404.2015.1018881

BOSCHMA R. and FRENKEN K. (2006) Why is economic geography not an evolutionary science? Towards an evolutionary economic geography, *Journal of Economic Geography* **6**, 273–302. doi:10.1093/jeg/lbi022

CAPELLO R. and LENZI C. (2014) Knowledge, innovation and productivity gains across European Regions, *Regional Studies*. doi:10.1080/00343404.2014.917167

DUSCHL M., SCHOLL T., BRENNER T., LUXEN D. and RASCHKE F. (2014) Industry-specific firm growth and agglomeration, *Regional Studies*. doi:10.1080/00343404.2015.1018881

EUROPEAN COMMISSION (2012) *Guide to Research and Innovation Strategies for Smart Specialization*. Publications Office of the European Union, Luxembourg.

FELDMAN M. (1994) *The Geography of Innovation, Economics of Science, Technology and Innovation*. Kluwer, Dordrecht.

FINGLETON B. and LOPEZ-BAZO E. (2006) Empirical growth models with spatial effects, *Papers in Regional Science* **85**, 177–198. doi:10.1111/j.1435-5957.2006.00074.x

FRATESI U. (2015) Regional knowledge flows and innovation policy: A dynamic representation, *Regional Studies*. doi:10.1080/00343404.2015.1068930

Galliano D., Magrini M.-B. and Triboulet P. (2014) Marshall's versus Jacobs' externalities in firm innovation performance: the case of French industry, *Regional Studies*. doi:10.1080/00343404.2014.950561

Guastella G. and Van Oort F. G. (2015) Regional heterogeneity and interregional research spillovers in European innovation: modelling and policy implications, *Regional Studies*. doi:10.1080/00343404.2015.1034668

McCann P. and Ortega-Argiles R. (2015) Smart specialization, regional growth and applications to European Union cohesion policy, *Regional Studies*. doi:10.1080/00343404.2013.799769

Regional Heterogeneity and Interregional Research Spillovers in European Innovation: Modelling and Policy Implications

GIANNI GUASTELLA† and FRANK G.VAN OORT‡§
†*Department of Mathematics and Physics, Università Cattolica, Brescia, Italy*
‡*Erasmus School of Economics Institute for Housing and Urban Development Studies (IHS), Erasmus University, Rotterdam, the Netherlands*
§*Department of Economic Geography, Faculty of Geosciences, Utrecht University, Utrecht, the Netherlands*

GUASTELLA G. and VAN OORT F. G. Regional heterogeneity and interregional research spillovers in European innovation: modelling and policy implications, *Regional Studies*. In agglomeration studies the effects of various regional externalities related to knowledge spillovers remain largely unclear. To explain innovation clustering, scholars emphasize the contribution of localized knowledge spillovers (LKS) and, specifically when estimating the knowledge production function (KPF), of (interregional) research spillovers. However, less attention is paid to other causes of spatial heterogeneity. In applied works, spatial association in data is econometrically related to evidence of research spillovers. This paper argues that, in a KPF setting, omitting spatial heterogeneity might lead to biased estimates of the effect of research spillovers. As an empirical test, a spatial KPF is estimated using EU-25 regional data, including a spatial trend to control for unexplained spatial variation in innovation. Accounting for geographical characteristics substantially weakens evidence of interregional research spillovers.

GUASTELLA G. and VAN OORT F. G. 欧洲创新中的区域异质性与区域间的研究外溢：模式化与政策意涵，*区域研究*。在聚集研究中，与知识外溢有关的各种区域外部性效应，仍然不甚明确。为了解释创新的集群作用，学者强调在地化的知识外溢（LKS）与（区域间的）研究外溢的贡献—特别是当评估知识生产功能（KPF）时。但却较少有研究关注空间异质性的其他肇因。在应用上，数据中的空间关联性，在计量经济方面，与研究外溢的证据有关。本文主张，在KPF 的环境中，忽略空间异质性，或许会导致对研究外溢效应的偏差评估。作为经验性的检验，本文将运用欧盟二十五座区域的数据来评估空间 KPF，包含控制创新中尚未被解释的空间变异之空间趋势。说明地理的特徵，将大幅弱化区域之间研究外溢的证据。

GUASTELLA G. et VAN OORT F. G. L'hétérogénéité régionale et les retombées de la recherche interrégionales en matière d'innovation européenne: la modélisation et les implications politiques, *Regional Studies*. Dans les études sur l'agglomération, les effets des diverses externalités régionales relatives aux retombées de l'application des connaissances restent en grande partie incertains. Pour expliquer le clustering de l'innovation, les chercheurs mettent l'accent sur la contribution des retombées localisées de l'application des connaissances et, notamment quand on estime la fonction de production de connaissances, des retombées de la recherche (interrégionales). Cependant, on prête moins d'attention aux autres causes de l'hétérogénéité spatiale. Dans la recherche appliquée, l'association spatiale des données se rapporte aux preuves des retombées de la recherche d'un point de vue économétrique. Cet article affirme que, dans le contexte de la fonction de production de connaissances, l'omission de l'hétérogénéité spatiale pourrait entraîner des estimations faussées de l'impact des retombées de la recherche. En guise d'un test empirique, on estime une fonction de production de connaissances spatiale employant des données régionales de l'EU à 25, y compris une tendance spatiale pour contrôler la variation spatiale inexpliquée de l'innovation. Tenir compte des caractéristiques géographiques affaiblit considérablement les preuves quant aux retombées de la recherche interrégionales.

GUASTELLA G. und VAN OORT F. G. Regionale Heterogenität und interregionale Forschungsübertragungen in der europäischen Innovation: Modellierung und politische Auswirkungen, *Regional Studies*. In Agglomerationsstudien bleiben die Effekte verschiedener regionaler Externalitäten im Zusammenhang mit Wissensübertragungen größtenteils unklar. Zur Erklärung der Clusterbildung im Bereich der Innovation betont die Wissenschaft den Beitrag der lokalisierten Wissensübertragungen sowie – speziell bei der Schätzung der Wissensproduktionsfunktion – von (interregionalen) Forschungsübertragungen. Den anderen Ursachen der räumlichen Heterogenität wird hingegen weniger Beachtung geschenkt. In angewandten Werken ist die räumliche Assoziation in Daten ökonometrisch mit den Belegen für Forschungsübertragungen verwandt. In diesem Beitrag wird argumentiert, dass ein Auslassen der räumlichen Heterogenität in Umgebungen der Wissensproduktionsfunktion zu verzerrten Schätzungen des Effekts von Forschungsübertragungen führen kann. Als empirischer Test wird eine räumliche Wissensproduktionsfunktion anhand von regionalen Daten der EU-25-Länder unter Berücksichtigung eines räumlichen Trends zur Kontrolle auf unerklärte räumliche Abweichungen bei der Innovation geschätzt. Bei einer Berücksichtigung der geografischen Merkmale schwächen sich die Belege für interregionale Forschungsübertragungen erheblich ab.

GUASTELLA G. y VAN OORT F. G. Heterogeneidad regional y transferencia de investigación interregional en la innovación europea: modelación y repercusiones políticas, *Regional Studies*. En los estudios de aglomeración, los efectos de las diferentes externalidades regionales relacionadas con la transferencia de conocimientos siguen siendo en gran medida poco claros. A fin de explicar las agrupaciones de innovación, los académicos destacan la contribución de la transferencia localizada de conocimientos y, en concreto al calcular la función de producción del conocimiento, de los efectos indirectos de la investigación (interregional). Sin embargo, se presta menos atención a otras causas de la heterogeneidad espacial. En los trabajos aplicados, la asociación espacial en datos está econométricamente relacionada con la evidencia de transferencia de investigación. En este artículo argumentamos que en un entorno de función de producción del conocimiento, al omitir la heterogeneidad espacial pueden obtenerse estimaciones sesgadas de la transferencia de investigación. Como prueba empírica, calculamos una función de producción del conocimiento espacial a partir de datos regionales de la UE-25, incluyendo una tendencia espacial para controlar la variación espacial no explicada en la innovación. Al tener en cuenta las características geográficas, se debilita en gran medida la evidencia de la transferencia de investigación interregional.

INTRODUCTION

The knowledge economy and innovation play central roles in numerous contemporary theories on economic development in Europe and policy debates. The European Union's ambition to be among the most competitive and innovative regions in the world translates to national and regional 'smart specializations' and place-based development strategies (BARCA *et al.*, 2012). The geographical co-location of innovative actors has been acknowledged as a key driver of regional competitiveness and academic research has devoted substantial attention to localized knowledge spillovers (LKS), in an attempt to explain the agglomeration and spatial concentration of innovative activities.

The relationship between agglomeration in cities and regions and innovation and economic development has been intensively studied in last decades. Since GLAESER *et al.* (1992), it has become common practice to analyse urban and regional growth variables using employment in cities and such efforts have suggested a relationship between agglomeration and economic growth, leading to the possibility that increasing returns to urbanization operate in a dynamic, rather than static, context. Sector-specific localization economies, stemming from input–output relationships and transport cost savings for firms, human capital externalities and knowledge spillovers, are generally compared with the formerly customary measures of general urbanization economies (HENDERSON, 2003). A substantial amount of literature builds on this conceptualization of agglomeration economies, which is reflected in three recent overviews and meta-studies (MELO *et al.*, 2009; BEAUDRY and SCHIFFAUEROVA, 2009; DE GROOT *et al.*, 2009). These studies reveal that the relationship between agglomeration and growth is ambiguous and inconclusive as to whether specialization or diversity is facilitated by (sheer) urbanization. This ambiguity is fuelled by measurement issues and heterogeneity in terms of temporal and spatial scales, aggregation, definitions of growth and the functional forms of the models applied. Based on an overview of historical and current conceptualizations of knowledge, knowledge diffusion and innovation in cities, several scholars call for conceptual and methodological renewal and rigour in future research to address this impasse in economic studies of agglomeration (VAN OORT and LAMBOOY, 2013). Only recently have various conceptualizations of distance and proximity been developed to address empirically the heterogeneity of the actors and processes involved and capture the role of cities and regions in this process. It is argued that research should more explicitly focus on both the transfer mechanisms of knowledge diffusion, such as spin-offs, research collaborations and social networks, and on the contexts that facilitate individual firms' familiarity with and diffusion of knowledge. New methodological perspectives also appear necessary, particularly modelling techniques that link appropriate levels of analysis, from the firm to regional contexts and agglomeration circumstances (VAN OORT *et al.*, 2012).

An important issue in much econometric research is the need to address regional heterogeneity in analyses of (knowledge) productive relationships (BASILE *et al.*, 2012b). This paper addresses this issue using knowledge production function (KPF) approaches that attempt econometrically to estimate and assess interregional research spillovers. The KPF is widely employed to investigate innovation dynamics in regions and spatial econometrics has substantially contributed to reshape the scope of research emphasizing the contributions of interregional innovation spillovers (ANSELIN *et al.*, 1997, 2000; PIERGIOVANNI and SANTARELLI, 2001; ACS *et al.*, 2002; BOTTAZZI and PERI, 2003; DEL BARRIO-CASTRO and GARCÍA-QUEVEDO, 2005; MORENO *et al.*, 2005; FRITSCH and SLAVTCHEV, 2007; PONDS *et al.*, 2010). The principal argument in this paper is that the majority of contributions in this line of research overemphasize the role of interregional innovation spillovers while failing to address optimally the role of markets and spatial heterogeneity, thereby leading to biased estimates of the magnitude of interregional spillovers. In this respect the paper contributes to the empirical literature on KPF in Europe by discussing the indirect nature and potential overvaluation of the evidence obtained by commonly applied modelling strategies (BRESCHI and LISSONI, 2001, 2009) and linking results to an integrative academic and policy debate. The paper demonstrates that spatial data models can address spatial heterogeneity in an intrinsic manner (rather than by relying on entirely different data sources, concepts or methods per se), thereby demonstrating its value in the burgeoning discussion of innovation spillovers in a regional context.

Using data on high-technology patenting activity in EU-25 NUTS-II (Nomenclature des Unités Territoriales Statistiques) regions in 2007–08, a KPF equation is estimated by employing a negative binomial model. The standard specification adopted in the empirical literature is extended to account for the role of markets in prompting regional innovation and local characteristics generally unobservable by the econometrician. To do so, a spatial trend is introduced in the estimated KPF as a smooth function of geographical coordinates as in BASILE *et al.* (2012a), and a semi-parametric generalized additive model (GAM) is estimated. The hypothesis is that when heterogeneity is not observed, estimates of the interregional research spillover effect are biased by the omission of geographical variables.

The evidence presented in this paper provides useful insights for academic and policy discussions and implications for regional policies. If spatial heterogeneity affects the desired treatment of individual networks and knowledge transfer mechanisms, varied local development strategies could be more appropriate than generalized ones. More importantly, if research externalities are only characterized by an interregional geographical scope to a limited extent, as the evidence in this paper suggests, then the co-location of innovative activities and agglomeration will likely only produce benefits in selected regions, causing disparities to increase. A recent policy discussion on place-based (BARCA, 2009; ORGANISATION FOR ECONOMIC CO-OPERATION AND DEVELOPMENT (OECD), 2009a, 2009b) versus place-neutral (WORLD BANK, 2009) development strategies in the European Union is highly relevant to the present topic and is summarized in BARCA *et al.* (2012). Place-neutral strategies rely on the agglomerative forces of the largest cities and metropolitan regions to attract talent and growth potential, further taking for granted the role of markets in the diffusion of the benefits of growth through the rest of the economy. Accordingly, agglomeration in combination with encouraging individual mobility not only allows individuals to live where they expect to be better off but also increases individual incomes, productivity, knowledge and aggregate growth (WORLD BANK, 2009). Advocates of place-based development strategies, in contrast, assert that the polycentric nature of a set of smaller and medium-sized cities in Europe, each with its own peculiar characteristics and specializing in the activities to which they are best suited, creates fruitful urban variety, which enhances optimal economic development, knowledge transfers and innovation. In fact, innovation and economic growth are not uniquely related to mega-city-regions (BARCA *et al.*, 2012) and the role of small and medium-size communities should be better addressed. This research confirms that innovation and economic growth are not uniquely related to mega-city-regions (BARCA *et al.*, 2012). Smart specialization, a policy tool for the division of money proposed for future European Union cohesion policies, could direct the focus of regional innovation opportunities (EUROPEAN COMMISSION, 2012).

The paper is organized as follows. The next section presents a critical review of the LKS theory. Criticisms of the theory are discussed and related to the limitations of empirical studies employing the KPF approach to address the role of interregional research spillovers. The empirical approach adopted in the analysis on European knowledge production is described in the third section. Data and results are introduced and discussed in the fourth section. The final section presents conclusions and connections to European innovation and cohesion policies.

INTERREGIONAL SPILLOVERS AND REGIONAL HETEROGENEITY IN THE KNOWLEDGE PRODUCTION FUNCTION (KPF)

The estimation of KPF at the territorial level dates back to JAFFE (1989). Its pervasiveness in the econometric literature on innovation dynamics is closely related to the ability to account for local externalities at the territorial level. Coefficient estimates of the relationship between patents and research and development (R&D) are found to be larger when the relationship is estimated at the territorial level due to local externalities in innovation. In particular, the co-location of universities and firms in a given region improves the effectiveness of R&D investments.

As knowledge and innovation are difficult to appropriate, their production often generates unintended benefits for agents (positive externalities) through several mechanisms. These benefits are deemed as 'localized' and hence are assessed as a cause of the geographical co-location of innovative activities. In the geography of innovation literature, particular attention is devoted to knowledge spillovers (AUDRETSCH and FELDMAN, 2004). Knowledge spillovers are broadly characterized as *pure* externalities and are related to the transfer of *tacit* knowledge between firms – and institutions – in non-market transactions. Understanding the geographical structures that underlie these spillover benefits is necessary for any evidence-based innovation policy to stimulate a region's (or collection of regions, such as Europe) transition towards a knowledge-based society. BOSCHMA and FRENKEN (2006) argue that tacit knowledge diffusion relies on specific transfer mechanisms such as labour mobility, spin-offs and inter-organizational networks, implying that the spillover effect is subject to distance decay. In recent years, numerous macro studies on the effect of knowledge spillovers on innovation have been conducted, and there is consensus that the strength of interregional knowledge flows decreases rapidly with geographical distance (ACS, 2002). The distance effect has been shown to exist even when spillovers are directly measured using patent citations (BRESCHI and LISSONI, 2009) using the approach initially proposed by JAFFE *et al.* (1993). As previous research has produced a certain degree of empirical agreement (FRITSCH and SLAVTCHEV, 2007), the *local* spatial dimension has been commonly acknowledged as a characteristic feature of such mechanisms. Recent applications at the regional and urban levels indicate that this line of reasoning is useful for explaining urban growth differentials (ACS, 2002). As a source of positive externalities, knowledge spillovers are theoretically capable of influencing the innovative behaviours of firms and, consequently, of determining regional disparities in the level of innovative activity and, eventually, economic performance (JAFFE *et al.*, 1993).

The approach is becoming increasingly sophisticated by incorporating interregional externalities conveyed through either the spatial structure or network structures of economies and scientists (MORENO *et al.*, 2005; BASILE *et al.*, 2012b; PONDS *et al.*, 2010; VARGA *et al.*, 2012). Geographical and relational externalities between regions are generally measured in an econometric sense, by estimating spatial lag and/or spatial error specifications, indicating potential spatial and relational diffusions of knowledge. This modelling framework is put forward incorporating the opportunity to include links between regions by using contiguity or other matrices that map relationships assumed to represent the infrastructure for the diffusion of externalities.

With the introduction of spatial econometric methods (ANSELIN, 1988a, 1988b), increased attention has been devoted to interregional innovation spillovers, examining the effect of R&D expenditures made in neighbouring regions on innovation. In an attempt to consider interregional externalities' contributions to innovation, the original cross-regional KPF, a linear relationship between innovative outputs and inputs, proxied by patents and R&D, respectively, has been extended by either including spatially lagged variables in the model or attributing a spatial structure to the error term. More precisely, the spatially lagged R&D variable is frequently included to account for the contribution of research performed in neighbouring regions (interregional spillovers in research). A substantial econometric literature based on spatial KPF has provided evidence supporting the existence of interregional innovation spillovers (ANSELIN *et al.*, 1997, 2000; PIERGIOVANNI and SANTARELLI, 2001; ACS *et al.*, 2002; DEL BARRIO-CASTRO and GARCÍA-QUEVEDO, 2005; FRITSCH and SLAVTCHEV, 2007). In addition, certain studies have more specifically investigated the geographical scope of knowledge spillovers in research by either testing the hypothesis at different distances (BOTTAZZI and PERI, 2003) or considering different distance bands (MORENO *et al.*, 2005). In a similar vein, this empirical framework has also been adapted to test the importance of spillovers mediated by technological proximity (GREUNZ, 2003) or institutional proximity, as proxied by scientific collaboration (PONDS *et al.*, 2010). Reduced geographical distance is necessary but not sufficient for knowledge diffusion (BOSCHMA, 2005). In general, there is econometrical evidence that innovation spillovers have a localized character.

Despite the large consensus achieved through the spatial econometric approach to the KPF, the identification of interregional innovation spillovers is liable of criticisms from both theoretical and empirical perspectives. From the theoretical perspective, the *pure* character of knowledge externalities cannot be easily identified (GEROSKI, 1995); however, this characteristic is frequently taken for granted in innovation studies, under the assumption that externalities originate from the informal transmission of tacit knowledge. Although informal knowledge exchange is very important for successful knowledge diffusion (DAHL and PEDERSEN, 2004), a consistent share of knowledge is generated through formal collaboration agreements between institutions (HAGEDOORN *et al.*, 2000) and unintended knowledge flows might only be complementary to the knowledge exchange mediated by market agreements. BRESCHI and LISSONI (2001), elaborating on these notions, suggest that the tacitness of knowledge is clearly not the only explanation for the clustering of innovative activities, as many studies on regional KPF seem to suggest. The presence of a market for technologies and specialized technology suppliers at the local level may promote spatial clustering among innovative firms potentially to a greater extent than pure knowledge spillovers (HENDERSON, 2003). A well-developed market for technology likely increases the market value of patents by aligning the demand for and supply of technologies, making it more convenient for innovative firms to locate within short distances from markets (LAMOREAUX and SOKOLOFF, 1999). Furthermore, the presence of specialized technology suppliers is expected to encourage the co-location of innovative firms, as a consequence of the resulting reduced complexity of innovative processes, provided that new technologies can be more easily acquired in the market than produced internally. Once innovative clusters begin to develop at the local level, this increases the likelihood that local firms will engage in R&D collaborations, and this might eventually produce spillovers. Accordingly, spillovers in research, both within the region and between regions, could be considered a consequence of the geographical co-location of firms rather than a cause.

Critiques have also been advanced concerning the empirical application of the KPF model. BOTTAZZI and PERI (2003) argue that coefficient estimates for the patent–research relationship are biased due to the omission of relevant variables that are highly related to both research investments and patenting activity. This is, for instance, the case for the market potential of a region, a variable which is likely to affect the R&D productivity of firms in the region and, consequently, their decision on the level of R&D investments. Market potential, a measure of the market available to firms in a region, is expected to be positively correlated with patents, as innovative firms might be willing to locate near to the market in which innovations are to be sold (VARGA *et al.*, 2012, term these the Edison-type of innovations). Similarly, a positive correlation can be expected between market potential and investments in research, as greater market opportunities indicate that a larger share of a firm's budget is committed to research. The findings reported in BOTTAZZI and PERI (2003) suggest that interregional innovation spillovers in research contribute little to regional innovation after accounting for R&D endogeneity.

Omitted variable bias and unobserved heterogeneity are both substantially important issues in the empirical estimation of the KPF and, to an even greater extent, when the model also accounts for interregional research spillovers. In summarizing the motivations for the use of a spatial econometric model, LESAGE and PACE (2009) highlight three central explanations for the evidence of spatial correlation in the data, namely omitted spatially correlated variables, unobserved spatial heterogeneity and externalities between units. Of these motivations, only the last resembles the argument for interregional innovation spillovers. In contrast, spatial autocorrelation in the model's residuals is likely related to the other two arguments in the case of a standard KPF. Variables commonly omitted from the model specification might exhibit a high degree of spatial association. This likely represents the case of market potential. Similarly, region-specific characteristics related to the demand for and supply of innovation are also frequently absent in the specification, due to their non-observability, causing unobserved spatial heterogeneity. Spatial model estimates do not allow the researcher to distinguish the spatial autocorrelation caused by unobserved spatial heterogeneity or omitted variables from that resulting from spatial interactions, and consequently it is possible to characterize as interregional knowledge spillovers what in reality is the effect of unobserved spatial heterogeneity and omitted variables.

The conceptual weaknesses of the LKS rationale for innovation clustering are closely related to the empirical issues arising in the spatially extended KPF. The mutual transfer of tacit knowledge via frequent interactions is clearly not the sole motivation for the spatial concentration of innovative activities. Market opportunities, on the one hand, determine regional innovation and attract innovative firms and stimulate the investments of existing firms; on the other hand, regional and local characteristics drive investments by providing an innovation-friendly environment for local firms (VARGA *et al.*, 2012). A failure to include these elements in the empirical specification produces an estimation bias that eventually results in incorrect inferences in favour of the LKS rationale. In support of this hypothesis, the results in TAPPEINER *et al.* (2008), based on 51 NUTS-I European Union regions, seem to indicate that evidence of spillovers is weakened by the inclusion of social capital variable in the model.

ECONOMETRIC STRATEGY

When estimated at the regional level, the KPF describes a linear relationship between patenting activity, a measure of the regional capacity to produce innovative output, and the R&D-to-GDP (gross domestic product) share, a measure of the innovative efforts made by firms and public institutions located in the region (JAFFE, 1989). This basic framework is extended by accounting for spatial relationships and spatial interactions between regions using spatial econometric techniques (ANSELIN *et al.*, 1997; ACS *et al.*, 2002; FISCHER and VARGA, 2003).

More recently, a number of studies – BOTTAZZI and PERI (2003) for EU-15 regions; FISCHER and VARGA (2003) for Austria; DEL BARRIO-CASTRO and

GARCÍA-QUEVEDO (2005) for Spain; FRITSCH and SLAVTCHEV (2005) for Germany; GUMBAU-ALBERT and MAUDOS (2009) for Spain; PONDS *et al.* (2010) for the Netherlands; AUTANT-BERNARD and LESAGE (2011) for France; and GRIMPE and PATUELLI (2011) for Germany – have concentrated on modelling the number of patents instead of the patenting rate (i.e., patents normalized per million inhabitants) in an attempt to maximize the informational content of the variable which is, by definition, discrete and positively defined. The analysis in this paper continues this line of research and, accordingly, distributions for count data are used to model the number of patent applications. In the majority of cases, interregional research spillovers are accounted for by including spatially lagged R&D; hence, the R&D variable is pre-multiplied by a row-standardized spatial weight matrix (DELTAS and KARKALAKOS, 2013).

For the set of 250 NUTS-II regions belonging to the EU-25 (Iceland and overseas territories excluded), regional innovation (PA_i) is measured as the average number of patent applications to the European Patent Office (EPO) during the period 2007–08. Applications are only considered if made in high-technology industries, in which the diffusion of tacit knowledge is expected to play a crucial role for innovation at both the firm and regional levels (KEEBLE and WILKINSON, 1999). More generally, evidence suggests that the diffusion of both formal and informal knowledge contributes to firm performance, especially in high-technology and science-based industries (PONDS, 2008). The definition adopted by EUROSTAT and based on International Patent Classification (IPC) classes is employed to classify patents in the high-tech category. The geographical distribution of total patents and patents in the high-tech sector is presented in Appendix A in the Supplementary data online alongside a more detailed description of the approach to patent classification.

The expected value of regional innovation is a function of (private) R&D expenditure by business enterprises (*BESRD*), university R&D (*UNIRD*) and government R&D (*GOVRD*), all relative to regional GDP. Private R&D investments primarily occur in regions with larger multinational enterprises, such as Eindhoven (Philips), Stockholm (Ericsson), Helsinki (Nokia), Leverkusen (Bayer), Stuttgart (Bosch, Porsche, Mercedes) and Toulouse (Airbus). University R&D is more associated with regions with technological universities and regions with alliance between universities and firms, such as Cambridge, Leiden, Braunschweig and Rome (DOGARU *et al.*, 2011). Additionally, the human capital available in the region, proxied by the share of regional employees with tertiary degrees (*TEREDUC*), is considered another input in the knowledge production process. Workforce education rather than tertiary education attainment data are used, speculating that the former best represents the actual employment of high-skilled workers in economic and innovation production. The contribution of interregional externalities in research is estimated by including the spatially lagged R&D to GDP ratio (*WBESRD*). These investments in innovation are hypothesized to have a positive correlation with innovation output, as much of the empirical literature also seems to suggest. The matrix W is defined as a positive definite weight matrix, the single element of which equals the squared inverse of the distance between regional centroids if this distance is less than 500 km; and zero otherwise.[1] The decision to rely on the 500 km cut-off is motivated by the observation that 492 km is the distance at which no region has no neighbours and, accordingly, each row of the matrix has at least one non-zero element. In addition, the share of employees in high- and medium/high-tech manufacturing (*HTMAN*) was included to control for sources of unobserved heterogeneity related to the industrial composition of the regional economy. The definition of Nomenclature statistique des activités économiques dans la Communauté européenne (NACE) sectors used to identify high- and medium/high-tech manufacturing industries is taken from EUROSTAT and is reported in Appendix A in the Supplementary data online. All right-hand-side variables are averaged over the period 2004–06 in an attempt to mitigate simultaneity bias. Finally, regional variation in economic size is captured by including a variable for average population for the period 2007–08 with a unity-constrained coefficient (offset). A complete description of the data set is provided in Appendix A.

Two indicators are included among explanatory variables to account for market factors that may influence the innovative activity of regions. Market potential (*MPOT*) is a measure of the size of the market that is potentially accessible from within the region. Data for this variable at the regional level are only available for 2006 and, in contrast to the other covariates, are not averaged over past years. The measure was created by the European Observation Network for Territorial Development and Cohesion (ESPON) project and is available for download at the project website.[2] Based on ESPON's definition, it is a proxy 'for the potential for activities and enterprises in the region to reach markets and activities in other regions'. Specifically, it is based on the average distance separating the region of origin from all other potentially accessible regions, and GDP values are used as weights. Accordingly, it is exogenous, while it is also appropriate to capture a measure of the potential markets of innovative firms, provided that distances are weighted by the respective GDP. In addition to the market potential indicator, an indicator of market size is included in the analysis, namely the gross value added per employee (*GVAPE*). The indicator is expected to account for the size of the internal market at the regional level and is, simultaneously, expected to explain variation in regional innovation related to the development of the region. As current values of the variable may induce endogeneity in the estimation, the value for 1991 is used, under the assumption that the lag is long enough to avoid any endogeneity.

Unobserved spatial heterogeneity is undoubtedly the most demanding of the specification problems. At the European regional level, the lack of sufficient regional data impedes the observation of relevant variables that likely affect the patenting activity of firms. In addition, many aspects influencing innovation in firms are not observable or measurable. This might be the case, for instance, when regional variation in innovation is related to the presence of regional markets for technologies and of specialized technology suppliers. Unobserved heterogeneity could be addressed by estimating model parameters using panel data techniques. It is plausible that most unobserved spatial heterogeneity would disappear after a within transformation. While longitudinal data on innovation are available for some European countries, this is unfortunately not the case for Europe as a whole. Although the patent database available at the EPO is among the most complete collections of data at the NUTS-II level, a complete longitudinal data set on R&D expenditures at the same level cannot be easily created due to limited data availability for certain regions. Accordingly, the use of longitudinal data would force the exclusion of certain regions from the data set.

In the absence of an available panel data set, the issue of unobservable heterogeneity can be addressed by including geographical variables. Unobservable regional characteristics are likely related to the business and institutional environments and hence non-randomly distributed in space. Among several alternatives, the inclusion of geographical dummy variables among the covariates would be the simplest and most intuitive solution. Nevertheless, this would require an ex ante division of the geographical space using a set of dichotomous variables, which, in turn, presumes knowledge on the way in which unobservable characteristics are distributed in space. Very common and easily interpretable choices include the use of country-specific dummy variables and, in the European case, dummy variables for regions in new member states and regions in cohesion countries. Particularly in the first case, research indicates that structural differences in economic environments may induce differentiated growth and innovation patterns (MARROCU *et al.*, 2012; DOGARU *et al.*, 2011).

A more direct approach to including unobserved spatial heterogeneity relies on the inclusion of regional geographical coordinates among the covariates. The relationship between innovation and geographical coordinates is expected to be non-linear, although the degree of non-linearity is unknown a priori. The linear hypothesis cannot be appropriately considered as long as it would imply that the number of patents increases (decreases) with increasing (decreasing) latitude or longitude. The most common non-linear specification, the quadratic hypothesis, can be considered more appropriate. Nonetheless, the specification is a only valid alternative when the spatial distribution of innovation exhibits a core–periphery pattern. The quadratic hypothesis suggests that innovation is increasing up to a certain threshold of longitude after which it begins decreasing. The same is expected to hold in case of latitude. Using parameter values to identify thresholds, it is possible to delineate a core area. Obviously, this impedes studying more complex patterns and might eventually result in misleading inferences in the latter case. To identify the functional form, this paper relies on the *spline* fitting method, on the basis of which the degree of non-linearity is selected by optimizing the informational content available in the data. Thus an isotropic thin plate regression spline, a non-linear function of joint longitude and latitude, is added to the linear predictor. The smoothed spatial trend is expected to account for the spatial variation in innovation explained by the geographical concentration of unobserved characteristics (BASILE *et al.*, 2012a).

The degree of non-linearity is selected by an algorithm minimizing:

$$\sum_i (y_i - f(x_i))^2 + \lambda \int f''(x)^2 \mathrm{d}x$$

where y is the dependent variable; and x is the set of independent variables (WOOD, 2006). The first term represents the residual sum of squares (RSS); the second is a penalty term based on the second derivative of the smooth function $f(\cdot)$. The penalty approaches zero as the smooth function becomes linear. Conversely, a high degree of non-linearity produces low RSS values but high penalty values. Therefore, the algorithm is demonstrably appropriate for weighting goodness of fit, on the one hand, and model complexity, on the other. In this respect, the choice of smoothed splines of geographical coordinates favours the more flexible non-linear specification for the spatial trend to the simple linear or quadratic specifications by correcting for the excess of non-linearity potentially caused by excessive unobserved heterogeneity.

The resulting model can be characterized as a generalized additive model (GAM) in which a non-parametric trend based on geographical coordinates *s*(long, lat) is added to a parametric (exponential) specification of the mean function. The estimation is performed using methodologies described by WOOD (2003, 2006) and BIVAND *et al.* (2008) and available in the R package *mgcv*. The final model is presented in equation (1). As explained, the dependent variable used in this model (*HTPC*) is obtained by averaging the values of patent applications over two consecutive years. No assumption is made about the distribution of the original variable. Instead, the appropriate model is selected relying on statistical tests conducted to explore which distribution best fits the dependent variable. Shapiro–Wilk normality tests were conducted, and the hypothesis that the distribution is normal or even log-normal is strongly rejected. As non-normality is likely the result of the high skewness of the distribution, an attempt was made to fit in the data up to the fourth quintile to the normal distribution, hence excluding the most patent-productive 20% of regions. While normality and log normality are also rejected in this case, the likelihood ratio statistic indicating the goodness of fit of the negative binomial distribution does not reject the null at a 5% significance level.[3] A negative binomial distribution is preferred to the more restrictive Poisson distribution, as the former relaxes the demanding assumption of mean-variance equality required by the latter with the introduction of the over-dispersion parameter θ. Furthermore, a plot of the Pearson residuals against fitted values of a Poisson distribution confirms that the variance increases with the expected value of the outcome variable, strengthening the preference for the negative binomial model against the Poisson model (a formal test for over-dispersion is also conducted and the result is presented in the next section). Of the 250 observations in the data set, only four regions contain zero applications. This suggests that the application of econometric procedures suitable to control for the abundance of zeroes is unnecessary.

$$\begin{aligned} HTPC_i &\sim \text{NB}(\mu_i, \theta) \\ \mu_i &= \exp\left(a + \sum_{k=1}^{K} b_k X_{k,i} + s(\text{long}_i, \text{lat}_i)\right) \end{aligned} \tag{1}$$

In the specification, the log link is used to relate the dependent variable to predictors. The exponential function[4] is a common specification for the mean function in generalized linear models (GLMs) in general, and in count data models in particular. Interpretation of coefficients is straightforward: holding constant other variables, a unit change in the predictor *k*th multiplies the number of patents by e^{b_k} (FOX, 2008, ch. 15). Predictors in the X matrix include the research variables (*BESRD*, *UNIRD* and *GOVRD*), controls (*HTMAN* and *TEREDUC*), market indicators (*MPOT* and *GVAPE*) and, finally, private research in neighbouring regions (*WBESRD*). All these effects are expected to be positive and significant. More specifically, a positive and significant coefficient on *BESRD* confirms the previous results regarding the validity of the KPF approach at the territorial level in Europe. Positive and significant estimates of coefficients related to *UNIRD* and *GOVRD* indicate the relevance of universities and government institutions, respectively, in shaping the geography of patenting activity in Europe. Evidence of interregional research spillovers is associated with an estimate of the *WBESRD* coefficient that is statistically greater than zero, as is standard in this literature (DELTAS and KARKALAKOS, 2013). Here it is argued that the evidence related to the last coefficient is biased if spatial heterogeneity is not taken into account. The magnitude and statistical significance of this coefficient is expected to decrease once the spatial trend is included in the specification. Dummy variables for regions in new member states and cohesion countries are also included in the regressions.

RESULTS

Table 1 provides a summary of the descriptive statistics for the variables in the model. By analysing the correlations, a positive relationship between patents and research conducted by private firms, universities and government institutions is detected, although to a lesser extent in the latter cases. All covariates exhibit a relatively low degree of correlation, both with patents and one another. A negative correlation is detected between the indicator of industrial composition (share of high- and medium/high-tech employment) and research in universities and human capital. This negative relationship may be due to the extent to which service industries and entrepreneurship are taken into account in the industrial composition indicator. VAN OORT and BOSMA (2013) find a similar negative relation in European regional data.

All average, minimum and maximum values suggest the absence of outliers in the data. As mentioned previously, the minimum value of applications equals to zero in certain – limited – cases, with a maximum of 706 and an average of 36. As the standard deviation substantially excides the mean, the mean-variance equality of the Poisson distribution may prove an unsuitable hypothesis in this case. On the contrary there is evidence of over-dispersion, which in turn suggests once again the use of a negative binomial distribution. The average R&D-to-GDP ratio is 0.89% in the case of expenditures by business enterprises, and 0.36% and 0.18% for university and government expenditures, respectively. Workers employed in high- and medium/high-tech manufacturing industries represent, on average, 6.39% of the total employment and 25% of workers hold a tertiary education degree.[5]

Turning to the spatial descriptive statistics, the Moran's Index I, computed on the basis of the 500 km inverse-squared distance weight matrix, is consistently positive, indicating spatial association for all variables. More specifically, large values of Moran's Index are reported for variables included to control for regional variation in high-tech patents not explained by R&D investments. In particular, the market potential and market size variables exhibit substantial spatial association and hence are expected to capture spatial heterogeneity to a large extent.

Before introducing geography into the econometric specification of the regional KPF, baseline results are presented and discussed in Table 2. Four models are

Table 1. Summary descriptive statistics of variables (EU-25 NUTS-II regions)

	HTPC	*BESRD*	*UNIRD*	*GOVRD*	*HTMAN*	*TEREDUC*	*MPOT*	*GVAPE*
Cross-correlations								
HTPC	–							
BESRD	0.5461	–						
UNIRD	0.2451	0.3286	–					
GOVRD	0.3273	0.3177	0.3813	–				
HTMAN	0.2302	0.4573	−0.0340	0.1050	–			
TEREDUC	0.2314	0.2877	0.2692	0.1910	−0.1190	–		
MPOT	0.3533	0.4258	0.2042	0.2845	0.3471	0.2978	–	
GVAPE	0.3675	0.4269	0.3331	0.0917	0.1205	0.3650	0.5346	–
Descriptive statistics								
Minimum	0.0000	0.0000	0.0000	0.0000	0.8170	13.3000	30.3000	1.9130
Mean	36.1800	0.8922	0.3615	0.1812	6.3900	25.1400	97.9300	32.8050
SD	76.4880	0.8870	0.2438	0.2025	3.4439	7.2848	35.2484	13.6234
Maximum	706.0000	4.9100	1.5900	1.1130	21.0730	50.2000	201.1000	64.4450
Moran test for spatial correlation								
I	0.0781	0.2679	0.1230	0.0846	0.3925	0.5367	0.6099	0.6387
E(I)	−0.0040	−0.0040	−0.0040	−0.0040	−0.0040	−0.0040	−0.0040	−0.0040
p-value	0.0000	0.0000	0.0000	0.0000	0.0000	0.0000	0.0000	0.0000

Note: p-values related to Moran's *I* are computed under a randomization and one-tailed hypothesis.

presented, a baseline model (a) that does not include market potential and market size indicators, a model in which the only market potential is included (b) and a model in which the market size is included jointly with market potential (c). Finally an attempt is made to mitigate the bias from unobserved heterogeneity by capturing country-specific effects through a set of dummy variables (d).

For all models, the results are also presented when the spatial lags of university research and human capital are included. Although the primary focus of this paper is the incidence of unobserved spatial heterogeneity in the estimation of private research spillovers, according to the geography of innovation literature, university research and the mobility of highly skilled workers are also important sources of knowledge spillovers that likely extend beyond regional borders (AUDRETSCH and FELDMAN, 2004). Accordingly, the spatial lags of both university research and tertiary education are included in all models to account for additional sources of interregional spillovers in innovation.

Baseline estimates of the regional KPF extended to account for interregional spillovers are presented in column (a). All coefficients related to research exhibit positive and highly significant values. The contribution of the share of high-tech employees in manufacturing is also positive and significant, as is the share of labour force with tertiary education. As expected, the coefficient associated with the dummy for regions in *NMS* is negative and significantly different from zero. The same result is found for the dummy for regions in *COHESION* countries. Concerning interregional spillovers, the spatial lag of private research is the only variable with a significant coefficient. Conversely, coefficients related to lagged university research and lagged human capital are both not significantly different from zero.

The value of θ is estimated at 2.082, and this value is consistently greater than 1 in all specifications. In addition to the difference between the mean and variance observed in the summary statistics for the dependent variable, this value of θ is considered clear evidence of over-dispersion. Nonetheless, for this specification only, a formal test of over-dispersion was conducted and definitively favours the negative binomial specification. The likelihood ratio statistic used to compare the Poisson with the negative binomial distribution takes a value of 3168, which is statistically significant at the 0.1% level.

The market potential indicator (*MPOT*) is added to the model in column (b). The inclusion of this variable leaves all previous result unchanged with the only notable exception being the *HTMAN* coefficient, which is now insignificant. The estimated coefficient related to market potential follows expectations and is positive and significant. The model in column (c) also includes gross value added per employee, a measure of market size (*GVAPE*). Although the market potential coefficient marginally decreases after the inclusion of market size, both coefficients are positively sloped and significant. Therefore, the two variables are able to account effectively for various aspects related to the potential and actual size of the regional market. As shown in Table 1, both the market potential and the market size variables exhibit very high degrees of spatial association and their omission from the model specification would likely result in substantial residual autocorrelation (MCMILLEN, 2003), which would eventually affect estimation. In the case of interregional innovation spillovers, the introduction of market

Table 2. Patent equation – generalized linear model (EU-25 NUTS-II regions)

	(a)	(b)	(c)	(d)
Intercept	−6.7577***	−7.3525***	−7.9210***	−7.2186***
	(0.3506)	(0.3744)	(0.4211)	(1.2443)
BESRD	0.5108***	0.4706***	0.4396***	0.4452***
	(0.0725)	(0.0701)	(0.0700)	(0.0732)
UNIRD	0.7400***	0.8144***	0.7412***	0.5430**
	(0.2368)	(0.2309)	(0.2281)	(0.2420)
GOVRD	0.7704***	0.6118**	0.8034***	0.7393***
	(0.2660)	(0.2587)	(0.2585)	(0.2673)
HTMAN	0.0408**	0.0231	0.0284	0.0140
	(0.0188)	(0.0186)	(0.0183)	(0.0211)
TEREDUC	0.0304**	0.0227*	0.0205*	0.0245**
	(0.0126)	(0.0124)	(0.0123)	(0.0121)
MPOT		0.0078***	0.0055***	0.0033
		(0.0018)	(0.0020)	(0.0023)
GVAPE			0.0204***	0.0165*
			(0.0079)	(0.0084)
NMS	−1.2866***	−1.0480***	−0.5096*	
	(0.1885)	(0.1923)	(0.2828)	
COHESION	−0.8535***	−0.4699**	−0.3881*	
	(0.1989)	(0.2155)	(0.2153)	
AT				0.4326
				(0.3540)
BE-NL-LU				0.3919
				(1.2300)
CZ-HU-SK-SI				−0.5769
				(1.2086)
DE-DK				0.4257
				(1.2114)
ES-PT				−0.6770
				(1.2279)
FR				0.1520
				(1.2235)
GR-CY				−1.1348
				(1.2568)
IT				−0.0467
				(1.2036)
PL-EE-LV-LT				−0.8462
				(1.2183)
SE-FI				0.4234
				(1.2654)
UK				−0.1694
				(1.2186)
W_BESRD	0.5463***	0.6416***	0.6096***	0.4032**
	(0.1629)	(0.1569)	(0.1548)	(0.1723)
W_UNIRD	−0.1563	0.2732	0.2985	−0.5145
	(0.4934)	(0.4983)	(0.4920)	(0.5918)
W_TEREDUC	0.0005	−0.0075	−0.0043	0.0010
	(0.0166)	(0.0162)	(0.0159)	(0.0183)
Theta	2.0820	2.2550	2.3300	2.5967
Akaike information criterion (AIC)	1758.5000	1742.0000	1736.7000	1733.8000
LZI	0.0686	0.0425	0.0380	0.0298
[*p*-value]	[0.0070]	[0.0571]	[0.0754]	[0.1259]

Notes: Standard error (SE) is given in parentheses. LZI is the value of the Moran's Index for residual autocorrelation adapted by LIN and ZHANG (2007) for GLM residuals. Since in all cases the LZI statistic exceeds the expected values, the one-side test for the 'greater' alternative hypothesis is conducted.

***Significance at 1% respectively; **at 5% respectively; and *at 10% respectively.

potential and market size affects the estimation of private research spillovers but not the significance level, while neither of the other two coefficients (on university research and tertiary education) becomes significant.

The results of LIN and ZHANG's (2007) test reject the null hypothesis of absence of spatial association in the residuals of models (a) (at a 5% significance level), (b) and (c) (but only at a 10% significance level). Notably, the degree of spatial association diminished after the introduction of market size and potential, indicating the effective contribution of the two variables in accounting for spatial heterogeneity. Nonetheless, unexplained spatial variation in the innovative activity of European Union regions still characterizes residuals, although to a narrowed extent only. Thus, country-specific dummy variables expected to capture unobserved heterogeneity at the country level are added to the model in column (d). Adopting a conservative approach in terms of degree of freedom, countries comprising either a single or a small number of NUTS-II regions have been aggregated. The variables *NMS* and *COHESION* were excluded to avoid perfect multicollinearity. Only small changes can be observed in the regression results, specifically concerning the effect of research in the region and neighbouring regions. The human capital variable continues to exhibit a positive and significant coefficient. In contrast, the market potential and market size coefficients both turn less significant, and even insignificant in the case of market potential. However, none of the dummy is statistically significant, and the utility of such a specification is thus questionable. To obtain a measure of the accuracy of the model, the estimates of the models in columns (c) and (d) were compared, using analysis of variance (ANOVA) analysis, with the null hypothesis of a model excluding both the *NMS* and *COHESION* dummy variables and country-specific dummy variables. As the model under the null hypothesis is nested in both alternatives, it is possible to compare alternatives. According to the results of the test, the restricted model is rejected in both cases but the associated *p*-value is lower in the case of model (c).[6] Based on this result, the model with dummy variables for new member states and cohesion countries is preferred to the model with country dummy variables, which, most important, may not lead to insightful policy results. Finally, evidence of spatial association in the residuals disappears in model (d).

Geographical information is included in the specification of the patent equation to account for unobserved spatial heterogeneity, and the results are presented in Table 3. A basic GLM is first estimated as a reference model (e). Based on the evidence in Table 2, dummy variables for new member states and cohesion countries are included instead of country-specific dummy variables and spatial lags of university research and human capital are removed because these are never significant. Comparing estimates of this baseline model in column (e) of Table 3 with the estimates in Table 2, it appears that the decision to simplify the specification is effective in terms of both parameter estimates and significance.

An initial attempt to introduce space in the regional KPF is made by adding geographical coordinates to the functional specification of the mean of patent counts. This is done in model (f), and related coefficients are reported. The coefficients for the two geographical variables are not significant. All other coefficients except for that associated to the dummy for cohesion countries are not affected by the change in the mean specification. Similarly, no significant changes emerge when the squared values of the coordinates are included in the specification. All coefficients, except that for cohesion countries, maintain their original signs, size and significance. Again, the coefficients for geographical variables are not significant.

The GAM estimates are reported in column (h). The test statistic concerning the significance of the spatial trend indicates that, in this case, geographical information significantly contributes to explaining variation in regional innovation. Coefficients related to private and university research continue to be highly significant and positive. The size of the coefficient associated with government research decreased and becomes less significant. Results indicate also a positive contribution of human capital to regional innovation and a positive relationship among high-tech patents, market potential and market size. Concerning the coefficient for research spillovers, its magnitude changes after the introduction of the spatial trend, and overall the coefficient is now only significant at the 10% level.

Fig. 1 maps the values of the estimated trend in the geographical space of European Union regions. The estimated value of the trend is indicated by a black line, while vertical and horizontal lines indicate the lower and upper bounds of the confidence level, respectively, set at ± 1 SE (standard error). In area below the lower bound (above the upper bound), the geographical position of the region contributes less (more) than expected to regional innovation, all other predictors being constant. Accordingly, it is unsurprising that regions in southern Italy and in certain *NMS* exhibit poor performance, while in contrast, regions in northern Italy, Germany and Denmark and Sweden perform very well. Values below the lower bound are also detected in eastern France. As this area is located between important technological poles in Europe (such as Eindhoven in the north and Toulouse in the south), this area performs worse than its geographical position would lead to expect. Finally, a similar situation is detected in the northern and eastern UK, although in this case there is no similar locational explanation for this evidence. This trend measures, to some extent, one's ignorance of the phenomenon and it might not be a simple matter to find a detailed explanation for the evidence.

The last part of Table 3 concerns the test for autocorrelation in the residuals. A LIN and ZHANG (2007) statistic applied to the residuals of models (e) to (g) indicates that the null hypothesis of absence of spatial association can be rejected at a 10% significance level. The low level of significance of the statistic shows that the residual autocorrelation is substantially mitigated by the inclusion of linear and squared geographical coordinates. Nonetheless, the null hypothesis of absence of spatial

Table 3. Patent equation – generalized linear and additive models (EU-25 NUTS-II regions)

	(e)	(f)	(g)	(h)
Intercept	−7.8680***	−8.4191***	−14.3600***	−7.8651***
	(0.3562)	(0.7231)	(4.4520)	(0.4184)
BESRD	0.4384***	0.4383***	0.4232***	0.4825***
	(0.0700)	(0.0699)	(0.0702)	(0.0683)
UNIRD	0.7439***	0.6444***	0.7413***	0.7528***
	(0.2272)	(0.2379)	(0.2410)	(0.2249)
GOVRD	0.8167***	0.8584***	0.8665***	0.4413*
	(0.2534)	(0.2575)	(0.2556)	(0.2500)
HTMAN	0.0280	0.0286	0.0268	0.0005
	(0.0180)	(0.0180)	(0.0186)	(0.0209)
TEREDUC	0.0189**	0.0165*	0.0153*	0.0293***
	(0.0083)	(0.0093)	(0.0092)	(0.0103)
MPOT	0.0053***	0.0056***	0.0051**	0.0063**
	(0.0019)	(0.0020)	(0.0022)	(0.0026)
GVAPE	0.0205***	0.0215***	0.0241***	0.0229**
	(0.0079)	(0.0080)	(0.0082)	(0.0091)
NMS	−0.5273*	−0.5971**	−0.7037**	−0.4228
	(0.2776)	(0.2919)	(0.2948)	(0.3374)
COHESION	−0.4201**	−0.3121	−0.2815	−0.7418*
	(0.2051)	(0.2396)	(0.2720)	(0.4239)
W_BESRD	0.6300***	0.5590***	0.5770***	0.3291*
	(0.1445)	(0.1576)	(0.1576)	(0.1850)
LONG		0.0046	−0.0074	
		(0.0074)	(0.0161)	
LAT		0.0121	0.2525	
		(0.0146)	(0.1781)	
$LONG^2$			0.0011	
			(0.0009)	
LAT^2			−0.0024	
			(0.0018)	
$s(LONG, LAT) - \chi^2$				51.7700
[*p*-value]				[0.0014]
Theta	2.3280	2.3500	2.3910	3.1310
Akaike information criterion (AIC)		1736.0000	1737.2000	1708.5940
LZI	0.0383	0.0360	0.0338	−0.0115
[*p*-value]	[0.0739]	[0.0853]	[0.0978]	[0.3992]

Notes: Standard error (SE) is given in parentheses. LZI is the value of the Moran's Index for residual autocorrelation adapted by LIN and ZHANG (2007) for GLM residuals. Significance is calculated based on the one-side alternative hypothesis. The selected alternative depends on whether the LZI statistic exceeds the expected value. The approximate significance of the *s*(*LONG, LAT*) term is evaluated using the χ^2-statistic test suggested by WOOD (2006). The value of the statistic is reported alongside the *p*-value associated with the null hypothesis of the term's statistical insignificance.

***Significance at 1% respectively; **at 5% respectively; and *at 10% respectively.

correlation in the residuals can only be rejected completely with the inclusion of the spatial trend.

The evidence in Table 3 indicates that the contribution of interregional research spillovers, measured on regional scale in Europe and while controlling for spatial heterogeneity, is limited. Prior research suggested that geographical spillovers have a positive effect on innovation. This effect is also observed in these results, but in conclusion, it appears less relevant than previously suggested. Overall, innovation in European regions is driven by several different factors, some of which are clearly unobservable by the econometrician. After including geographical coordinates, in an attempt to control for geographical heterogeneity not captured by other covariates, evidence of interregional research spillovers is weakened considerably. This indicates evidence of spatial correlation that is caused, in this case, by model misspecification, the omission of relevant variables and unobserved spatial heterogeneity, thereby assigning only a secondary role to geographically LKS.

Robustness checks

In the remainder of this section, the validity of this result is assessed by comparing GLM (without trend) estimates with GAM (with trend) counterpart estimates against changes in the specification of the spatial weight matrix used to construct the spatial lag of private research expenditures.

Results are reported in Appendix A in the Supplementary data online for four values of the critical

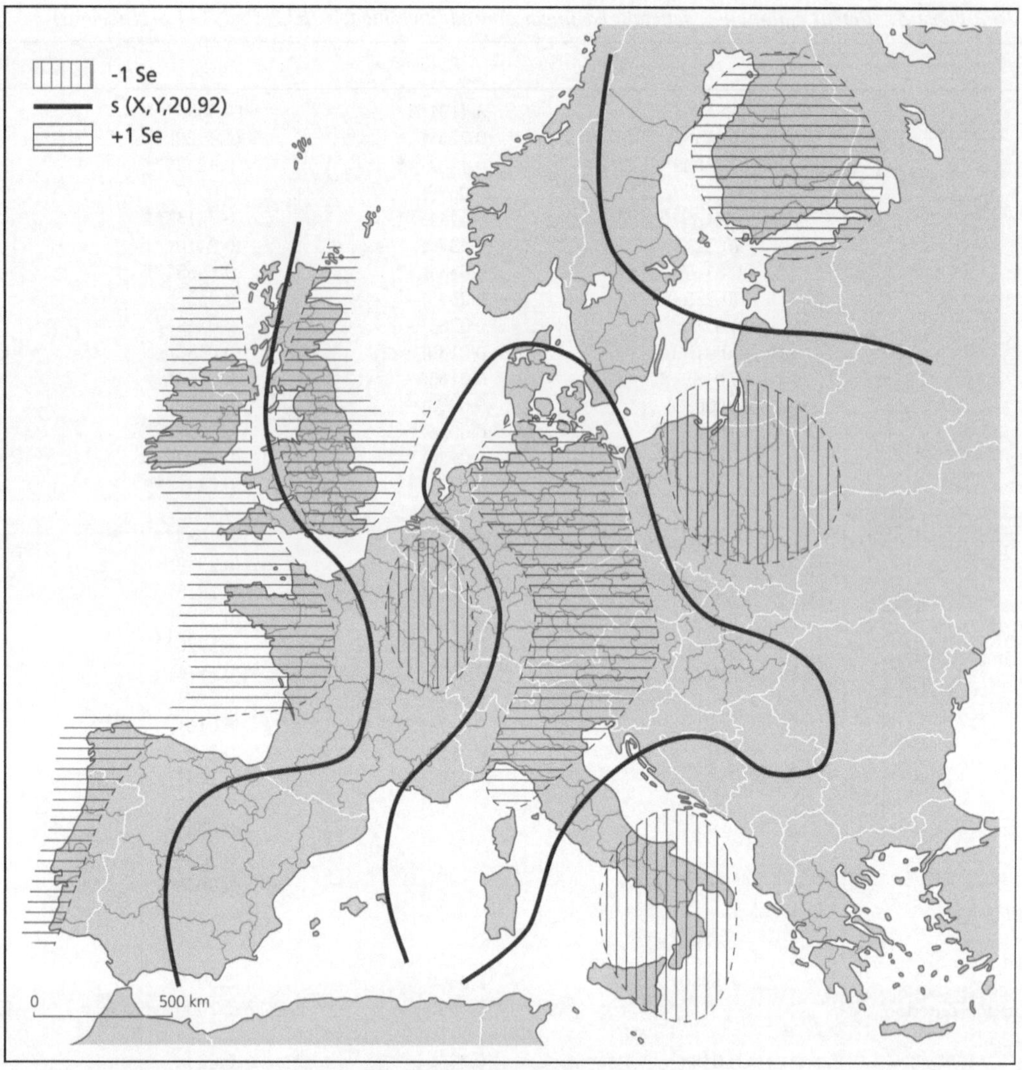

Fig. 1. Estimated spatial trend

distance used to construct the contiguity matrix. Weights are then computed consistent with the previous definition; hence the inverse of squared distance among centroids is used as a weight and row-standardization is applied.

All coefficients related to research have the expected sign and are statistically significant. The change in the critical cut-off distance does not affect the size or the significance level of the estimated coefficients. A change in significance is only observed in the case of the share of the *HTMAN* variable, which, consistent with previous evidence, is not significant when a value of the cut-off distance greater than 500 km is used. Conversely, coefficients related to human capital, market potential and market access, and the two dummies for new member states and cohesion countries, exhibit the correct signs and high significance levels for all values of the critical distance. Finally, the results indicate, at least for those of GLM estimates, the positive contribution of interregional research spillovers to regional innovation. In addition, the estimated effect of research in neighbouring regions is not constant becomes larger at higher values of the critical distance.

By focusing on the model specification, residual autocorrelation is detected in all models, providing confirmation that dummy variables, market potential and market access may not be sufficiently able to account for spatial heterogeneity.

Comparing GLM and GAM estimates, the latter including the smooth trend, the only significant change in coefficient estimates concerns the government R&D variable. As expected, the two geographical dummy variables for new member states and cohesion countries are no longer significant. Focusing on the variable of interest, interregional research spillovers, the associated coefficient is lower in magnitude and less significant in all models with the spatial trend. The difference between the GLM and GAM estimates is large and evident in all models, independent of the critical cut-off value. For distance values below 500 km, the coefficient becomes insignificant after the inclusion of the trend, while for values greater than 500 km it is only weakly significant. All tests on the GAM residuals fail to reject the null of absence of spatial correlation, while the χ^2 statistic related to the spatial trend is always significant.

To conclude, changes in the critical cut-off distance used to construct the weight matrix do not significantly affect the main evidence in this paper, as evidence of interregional research spillovers appears less relevant for regional innovation when spatial heterogeneity is accounted for in the model specification.

CONCLUSIONS AND DISCUSSION

Research spillovers have received increasing attention in the empirical literature on regional innovation, as they have been indicated as one of the most important vehicles for regional growth. Studies seeking evidence of research spillovers have argued that such spillovers are localized, as tacit knowledge cannot be codified and requires face-to-face contacts to be exchanged. Using a spatially extended KPF, this paper examined the contribution of interregional research spillovers to regional innovation in the European Union, with particular focus on spillovers between neighbouring regions. This paper argued that when studying the spatial clustering of innovative activities, past research has overemphasized the LKS argument both theoretically and empirically, and less attention has been devoted to the geographical characteristics of the region. This emphasis on innovation spillovers has been empirically supported by spurious results, which are often biased by the omission of geographical variables from the econometric specification. At the regional aggregate level, innovative output is not caused by R&D alone. Rather, output relates to region-specific characteristics such as the industry mix, market opportunities, the innovative environment and social capital. Some of these characteristics can be observed, while others cannot. Each, however, also relates to R&D to the extent that it affects the productivity of R&D investments and, consequently, to the R&D investment decision of local firms. Their omission from the econometric specification introduces bias in the R&D-related coefficients. Finally, if these variables have a specific spatial structure, similar to that of R&D investments, their omission is also likely to bias the lagged R&D coefficient, causing incorrect inferences regarding interregional research spillovers.

The evidence provided in this paper nuances and, to some extent, contradicts results presented in previous empirical contributions. However the analysis does not contradict LKS theory in general or the potential relevance of knowledge spillovers in particular. Rather, this paper suggested that methods currently employed to address the issue of interregional spillovers in research collaborations risk-producing biased results. This, in turn, requires researchers to improve the conceptualization and econometric specification of the KPF, including the introduction of multilevel interaction scales, ranging from the very local to the regional, as well as network measures of spatial interaction and the specific multilevel interaction. Knowledge spillovers are likely to be either highly localized, and hence evidence of interregional spillovers should be sought at a low geographical scale, or captured in (social and cooperation) network relationships. Assessing this in specific technology fields, in panel data settings and controlling for additional regional heterogeneity variables is essential.

The results presented in this paper have important policy implications. It appears that excessive emphasis has likely been placed on the local knowledge spillover theory in explaining spatial patterns of innovation, and attention should be refocused on what actually motivates innovative investments. At the regional aggregate level, the evidence suggests that innovation is led by existing market opportunities and regional innovative environments. Spillovers are thus likely also the consequence of innovation clustering, rather than merely the cause. Findings in this paper therefore also contribute to the recent discussion on place-based or place-neutral development strategies in the European Union (BARCA *et al.*, 2012). The conceptual discussion of development oscillates between, on the one hand, spatially blind approaches arguing that intervention without respect to context ('people-based policy') is the best approach and, on the other hand, place-based approaches assuming that interactions between institutions and geography are critical for development. Research results discussed in this paper indicate a substantial heterogeneity in regional and urban conditions that influence patenting activity in Europe, suggesting that micro-economic processes and network alliances operate differently in different regions. This supports European place-based policy strategies *alongside* place-neutral (people-based) policy strategies. Both types of strategies are important for innovation policies intended to promote research cooperation and dissemination. Clearly, this research indicates that a more careful analysis of individual and network-level innovation processes in a multilevel spatial framework is needed to capture the full advantages of the two policy strategies.

Acknowledgments – The final version of this paper benefitted from the many useful comments and suggestions received at the VIth conference of the Spatial Econometric Association as well as at the first Geography of Innovation seminar. The authors would also like to thank two anonymous referees for providing good suggestions. The usual disclaimer applies.

Disclosure statement – No potential conflict of interest was reported by the authors.

Supplemental data – Supplemental data for this article can be accessed at http://dx.doi.org/10.1080/00343404.2015.1034668

NOTES

1. As usual, the matrix is row standardized.
2. See http://www.espon.eu/.
3. The variable is rounded to the integer. Detailed results of these tests are not presented but are available from the authors upon request.
4. The mean function is the inverse of the link function.
5. The market potential variable has an average value of 97.93. This differs from the value of 100, which is expected because the indicator is measured as the deviance from the European Union mean (EU = 100), as the Atlantic Islands were not considered in the data set used for the analysis.
6. Specifically, the *p*-value of the ANOVA test when model (c) is considered as the alternative is 4.2188E–15, while when model (d) is considered, the alternative the *p*-value decreases to 0.2054E–15.

REFERENCES

Acs Z. J. (2002) *Innovation and the Growth of Cities*. Edward Elgar, Cheltenham.

Acs Z. J., Anselin L. and Varga A. (2002) Patents and innovation counts as measure of regional production of new knowledge, *Research Policy* **31**, 1069–1085.

Anselin L. (1988a) Lagrange multiplier test diagnostic for spatial dependence and spatial heterogeneity, *Geographical Analysis* **20**, 1–17. doi:10.1111/j.1538-4632.1988.tb00159.x

Anselin L. (1988b) *Spatial Econometrics Methods and Models*. Kluwer, Boston, MA.

Anselin L., Varga A. and Acs Z. (1997) Local geographical spillovers between university research and high technology Innovations, *Journal of Urban Economics* **42**, 422–448. doi:10.1006/juec.1997.2032

Anselin L., Varga A. and Acs Z. (2000) Geographical spillovers and university research: a spatial econometric perspective, *Growth and Change* **31**, 501–515. doi:10.1111/0017-4815.00142

Audretsch D. B. and Feldman M. P. (2004) Knowledge spillovers and the geography of innovation, in Henderson J. V. and Thiesse J. F. (Eds) *Handbook of Regional and Urban Economics*, pp. 2713–2739. Elsevier/North Holland, Amsterdam.

Autant-Bernard C. and LeSage J. P. (2011) Quantifying knowledge spillovers using spatial econometric models, *Journal of Regional Science* **51**, 471–496. doi:10.1111/j.1467-9787.2010.00705.x

Barca F. (2009) *An Agenda for a Reformed Cohesion Policy: A Place-Based Approach to Meeting European Union Challenges and Expectations*. Report for the European Commission, Brussels.

Barca F., McCann P. and Rodriguez-Pose A. (2012) The case for regional development intervention: place-based versus place-neutral approaches, *Journal of Regional Science* **52**, 134–152. doi:10.1111/j.1467-9787.2011.00756.x

Basile R., Benfratello L. and Castellani D. (2012a) Geoadditive models for regional count data: an application to industrial location, *Geographical Analysis* **45**, 28–48. doi:10.1111/gean.12001

Basile R., Capello R. and Caragliu A. (2012b) Technological interdependence and regional growth in Europe: proximity and synergy in knowledge spillovers, *Papers in Regional Science* **91**, 697–722.

Beaudry C. and Schiffauerova A. (2009) Who's right, Marshall or Jacobs? The localization versus urbanization debate, *Research Policy* **38**, 318–337.

Bivand R., Pebesma E. J. and Gómez Rubio V. (2008) *Applied Spatial Data Analysis with R*. Springer, Berlin.

Boschma R. (2005) Proximity and innovation: a critical assessment, *Regional Studies* **39**, 61–74. doi:10.1080/0034340052000320887

Boschma R. and Frenken K. (2006) Why is economic geography not an evolutionary science? Toward an evolutionary economic geography, *Journal of Economic Geography* **6**, 273–302. doi:10.1093/jeg/lbi022

Bottazzi L. and Peri G. (2003) Innovation and spillovers in regions: evidence from European patent data, *European Economic Review* **47**, 687–710. doi:10.1016/S0014-2921(02)00307-0

Breschi S. and Lissoni F. (2001) Knowledge spillovers and local innovation systems: a critical survey, *Industrial and Corporate Change* **10**, 975–1005. doi:10.1093/icc/10.4.975

Breschi S. and Lissoni F. (2009) Mobility of skilled workers and co-invention networks: an anatomy of localized knowledge flows, *Journal of Economic Geography* **9**, 439–468. doi:10.1093/jeg/lbp008

Dahl M. S. and Pedersen C. O. R. (2004) Knowledge flows through informal contacts in industrial clusters: myth or reality?, *Research Policy* **33**, 1673–1686. doi:10.1016/j.respol.2004.10.004

De Groot H., Poot J. and Smit M. J. (2009) Agglomeration externalities, innovation and regional growth: theoretical perspectives and meta-analysis, in Nijkamp P. and Capello R. (Eds) *Handbook of Regional Growth and Development Theories*, pp. 256–281. Edward Elgar, Cheltenham.

Del Barrio-Castro T. and García-Quevedo J. (2005) Effects of university research on the geography of innovation, *Regional Studies* **39**, 1217–1229. doi:10.1080/00343400500389992

Deltas G. and Karkalakos S. (2013) Similarity of R&D activities, physical proximity and R&D spillovers, *Regional Science and Urban Economics* **43**, 124–131. doi:10.1016/j.regsciurbeco.2012.06.002

Dogaru T., Van Oort F. and Thissen M. (2011) Agglomeration economies in European regions: perspectives for objective-1 regions, *Tijdschrift voor Economische and Sociale Geografie* **102**, 486–494. doi:10.1111/j.1467-9663.2011.00676.x

European Commission (2012) *Guide to Research and Innovation Strategies for Smart Specialisation*. European Commission, Brussels.

Fischer M. and Varga A. (2003) Spatial knowledge spillovers and university research: evidence from Austria, *Annals of Regional Science* **37**, 303–322. doi:10.1007/s001680200115

Fox J. (2008) *Applied Regression Analysis and Generalized Linear Models*. Sage, Thousand Oaks, CA.

Fritsch M. and Slavtchev V. (2005) *The Role of Regional Knowledge Sources for Innovation? An Empirical Assessment*. No. 2005, 15. Freiberger Arbeitspapiere, 2005. Technische Universität Bergakademie Freiberg, Freiberg.

Fritsch M. and Slavtchev V. (2007) Universities and Innovation in Space, *Industry and Innovation* **14**, 201–218. doi:10.1080/13662710701253466

Geroski P. (1995) Markets for technology: knowledge, innovation and appropriability, in Stoneman P. (Ed.) *Handbook of Economics of Innovation and Technological Change*, pp. 90–131. Blackwell, Oxford.

Glaeser E. L., Kallal H. D., Scheinkman J. A. and Schleifer A. (1992) Growth in cities, *Journal of Political Economy* **100**, 1126–1152. doi:10.1086/261856

Greunz L. (2003) Geographically and technologically mediated knowledge spillovers between European regions, *Annals of Regional Science* **37**, 657–680. doi:10.1007/s00168-003-0131-3

Grimpe C. and Patuelli R. (2011) Regional knowledge production in nanomaterials: a spatial filtering approach, *Annals of Regional Science* **46**, 519–541. doi:10.1007/s00168-009-0355-y

Gumbau-Albert M. and Maudos J. (2009) Patents, technological inputs and spillovers among regions, *Applied Economics* **41**, 1473–1486.

Hagedoorn J., Link A. N. and Vonortas N. S. (2000) Research partnerships, *Research Policy* **29**, 567–586. doi:10.1016/S0048-7333(99)00090-6

Henderson J. V. (2003) Marshall's scale economies, *Journal of Urban Economics* **53**, 1–28. doi:10.1016/S0094-1190(02)00505-3

Jaffe A. B. (1989) Real effects of academic research, *American Economic Review* **79**, 957–970.

Jaffe A. B., Trajtenberg M. and Henderson R. (1993) Geographic localization of knowledge spillover as evidence by patent citations, *Quarterly Journal of Economics* **108**, 577–598. doi:10.2307/2118401

Keeble D. and Wilkinson F. (1999) Collective learning and knowledge development in the evolution of regional clusters of high technology SMEs in Europe, *Regional Studies* **33**, 295–303. doi:10.1080/00343409950081167

Lamoreaux N. R. and Sokoloff K. L. (1999) Inventors, firms, and the market for technology in the late nineteenth and early twentieth centuries, in Lamoreaux N. R., Raff D. M. G. and Temin P. (Eds) *Learning by Doing in Markets, Firms, and Countries*, pp. 19–60. University of Chicago Press, Chicago, IL.

LeSage J. and Pace K. (2009) *Introduction to Spatial Econometrics*. CRC Press, Boca Raton, FL.

Lin G. and Zhang T. (2007) Loglinear residual tests of Moran's *I* autocorrelation and their applications to Kentucky breast cancer data, *Geographical Analysis* **39**, 293–310. doi:10.1111/j.1538-4632.2007.00705.x

Marrocu E., Paci R. and Usai S. (2012) Productivity growth in the old and new Europe: the role of agglomeration externalities, *Journal of Regional Science* **53**, 418–442.

McMillen D. P. (2003) Spatial autocorrelation or model misspecification?, *International Regional Science Review* **26**, 208–217. doi:10.1177/0160017602250977

Melo P. C., Graham D. J. and Noland R. B. (2009) A meta-analysis of estimates of urban agglomeration economies, *Regional Science and Urban Economics* **39**, 332–342. doi:10.1016/j.regsciurbeco.2008.12.002

Moreno R., Paci R. and Usai S. (2005) Spatial spillovers and innovation activity in European regions, *Environment and Planning* **37**, 1793–1812. doi:10.1068/a37341

Organisation for Economic Co-operation and Development (OECD) (2009a) *How Regions Grow*. OECD, Paris.

Organisation for Economic Co-operation and Development (OECD) (2009b) *Regions Matter: Economic Recovery, Innovation and Sustainable Growth*. OECD, Paris.

Piergiovanni R. and Santarelli E. (2001) Patents and the geographic localization of R&D spillovers in French manufacturing, *Regional Studies* **35**, 697–702.

Ponds R. (2008) Regional Innovation and the geography of research collaboration in science-based industries. PhD dissertation, Faculty of Geosciences, Utrecht University.

Ponds R., Van Oort F. and Frenken K. (2010) Innovation, spillovers and university–industry collaboration: an extended knowledge production function approach, *Journal of Economic Geography* **10**, 231–255.

Tappeiner G., Hauser C. and Walde J. (2008) Regional knowledge spillovers: fact or artifact?, *Research Policy* **37**, 861–874. doi:10.1016/j.respol.2007.07.013

Van Oort F. and Bosma N. (2013) Agglomeration economies, inventors and entrepreneurs as engines of European regional productivity, *Annals of Regional Science* **51**, 213–244.

Van Oort F., Burger M., Knoben J. and Raspe O. (2012) Multilevel approaches and the firm-agglomeration ambiguity in economic growth studies, *Journal of Economic Surveys* **26**, 468–491. doi:10.1111/j.1467-6419.2012.00723.x

Van Oort F. and Lambooy J. (2013) Cities, knowledge and innovation, in Fischer M. and Nijkamp P. (Eds) *Handbook of Regional Science*, pp. 475–488. Springer, Berlin.

Varga A., Pontikakis D. and Chorafakis G. (2012) Metropolitan Edison and cosmopolitan Pasteur? Agglomeration and interregional research network effects on European R&D productivity, *Journal of Economic Geography* **14**, 229–263. doi:10.1093/jeg/lbs041

Wood S. (2006) *Generalized Additive Models: An Introduction with R*. Chapman & Hall/CRC Press, Boca Raton, FL.

Wood S. N. (2003) Thin plate regression splines, *Journal of the Royal Statistical Society* **65**, 95–114. doi:10.1111/1467-9868.00374

World Bank (2009) *World Development Report 2009: Reshaping Economic Geography*. World Bank, Washington, DC.

Knowledge, Innovation and Productivity Gains across European Regions

ROBERTA CAPELLO and CAMILLA LENZI
Department ABC, Politecnico di Milano, Milan, Italy

CAPELLO R. and LENZI C. Knowledge, innovation and productivity gains across European regions, *Regional Studies.* This paper studies the relationship between knowledge, innovation and productivity in European regions, looking for the sources of spatial heterogeneity across regions in the type of knowledge needed for innovation and in the propensity to innovate, which mostly depend upon specific and systemic context conditions. The conceptual framework of analysis is based on the strong belief that different modes of innovation characterize regions, according to the presence/absence of some context conditions that allow for the creation and/or adoption of knowledge and innovation. Given the differences in knowledge and innovation intensity that characterize each mode/pattern of innovation, expectations are formulated on the achievement of productivity gains in regions characterized by different innovation patterns. Empirical evidence supports the conceptual expectations and shows that in regional innovation patterns based on local scientific knowledge-creation processes there are positive returns to scientific knowledge. However, in less knowledge-intensive patterns considerable productivity gains can also be achieved by local actors exploiting non-scientific knowledge and innovative capacity. Importantly, whereas the benefits accruing from knowledge appear rather selective and concentrated in a relatively small number of regions, the benefits generated by innovation seem more pervasive and beneficial, even in regions with a low endowment of scientific knowledge. These results have considerable implications for the current design of European Union innovation policies at the regional level.

CAPELLO R. and LENZI C. 欧盟各区域的知识、创新与生产力增加，*区域研究*。本文研究欧洲区域中，知识、创新与生产力之间的关联性，寻求各区域之间就创新所需的知识类型和创新倾向上的空间异质性来源，而它们多半取决于特定且系统性的脉络条件。分析的概念架构，来自于深信区域根据考量创造及/或採用知识与创新的脉络条件的部分缺少或存在，而具有不同的创新模式。考量各创新模式/形式的知识与创新密度的差异，形塑对于具有不同创新形式的区域中的生产力增加的期待。经验证据支持概念化的期待，并显示出，在基于在地科学知识创新过程的区域创新形式中，对科学知识有着正向回馈。但在知识密度较低的形式中，亦可透过在地行动者运用非科学性的知识和创新能力，获得大量的生产力增加。重要的是，当从知识获取的益处呈现选择性、且聚集于相对少数的区域时，创新所生产的益处似乎更加普遍且有益，即便是在拥有低度科学知识的区域之中亦是如此。这些研究结果，对于欧盟目前在区域层级的创新政策设计而言，具有相当的意涵。

CAPELLO R. et LENZI C. La connaissance, l'innovation et les gains de productivité à travers les régions européennes, *Regional Studies.* Cet article étudie le rapport entre la connaissance, l'innovation et les gains de productivité des régions européennes, recherchant les origines de l'hétérogénéité spatiale à travers les régions quant au type de connaissance nécessaire à l'innovation et pour ce qui est de la propension à innover, lesquels dépendent dans une large mesure des conditions de contexte à la fois spécifiques et systémiques. Le cadre conceptuel de l'analyse repose sur la ferme conviction que les régions se caractérisent par des modes d'innovation différents, selon la présence/l'absence de certaines conditions de contexte qui permettent la création et/ou l'adoption de la connaissance et de l'innovation. Étant donné les différences de l'intensité de la connaissance et de l'innovation qui caractérisent chaque mode/schéma d'innovation, on élabore des attentes en termes de la réalisation des gains de productivité dans les régions caractérisées par des schémas d'innovation différents. Des preuves empiriques soutiennent les attentes conceptuelles et montrent que les schémas d'innovation régionaux fondés sur les processus de production de la connaissance scientifique locale apportent des rendements positifs à la connaissance scientifique. Cependant, pour les schémas à faible intensité de connaissance, d'importants gains de productivité sont aussi à réaliser par les agents locaux qui exploitent la connaissance et la capacité d'innovation non-scientifiques. Il est à noter que les avantages dus à l'innovation semblent plus omniprésents et bénéfiques, même dans les régions dont la dotation en connaissance scientifique s'avère peu élevée, tandis que les avantages qu'entraîne la connaissance semblent plutôt sélectifs et se concentrent dans une poignée de régions. Ces résultats ont d'importantes retombées pour la conception actuelle des politiques d'innovation de l'Union européenne sur le plan régional.

CAPELLO R. und LENZI C. Wissen, Innovation und Produktivitätszuwachs in europäischen Regionen, *Regional Studies*. In diesem Beitrag untersuchen wir die Beziehung zwischen Wissen, Innovation und Produktivität in europäischen Regionen unter besonderer Berücksichtigung der Quellen von räumlicher Heterogenität in den Regionen hinsichtlich der für Innovation benötigten Art von Wissen sowie hinsichtlich der Innovationsneigung, die größtenteils von spezifischen und systemischen Kontextbedingungen abhängt. Der konzeptuelle Rahmen der Analyse beruht auf der festen Überzeugung, dass sich Regionen durch verschiedene Modi von Innovation auszeichnen, was von der Präsenz bzw. Abwesenheit bestimmter Kontextbedingungen abhängt, welche die Erzeugung und/oder Anpassung von Wissen und Innovation ermöglichen. Angesichts der Unterschiede bei der Intensität von Wissen und Innovation, die die verschiedenen Modi bzw. Muster der Innovation auszeichnen, werden Erwartungen hinsichtlich des Erreichens von Produktivitätszuwächsen in Regionen mit unterschiedlichen Innovationsmustern formuliert. Die empirischen Belege bekräftigen die konzeptuellen Erwartungen und zeigen, dass in regionalen Innovationsmustern, die auf Prozessen zur Erzeugung von lokalen wissenschaftlichen Erkenntnissen beruhen, positive Ergebnisse hinsichtlich der wissenschaftlichen Erkenntnisse zu verzeichnen sind. In weniger wissensintensiven Mustern lassen sich jedoch ebenfalls erhebliche Produktivitätssteigerungen erzielen, wenn lokale Akteure nichtwissenschaftliche Erkenntnisse und innovative Kapazität nutzen. Wichtig ist hierbei, dass der aus Wissen erwachsende Nutzen recht selektiv und auf eine relativ kleine Anzahl von Regionen konzentriert zu sein scheint, während der aus Innovation erwachsende Nutzen allgegenwärtiger und selbst in Regionen mit einem geringen Maß von wissenschaftlichen Erkenntnissen nützlicher erscheint. Diese Ergebnisse sind für die derzeitige Gestaltung der Innovationspolitik der Europäischen Union auf regionaler Ebene von erheblicher Bedeutung.

CAPELLO R. y LENZI C. Conocimiento, innovación y aumento de la productividad en las regiones europeas, *Regional Studies*. En este artículo analizamos la relación entre el conocimiento, la innovación y la productividad en las regiones europeas, teniendo en cuenta las fuentes de la heterogeneidad espacial en las diferentes regiones según el tipo de conocimiento necesario para la innovación y según la predisposición a innovar, que depende mayoritariamente de condiciones de contexto específicas y sistémicas. La estructura conceptual del análisis se basa en la opinión arraigada de que las regiones se caracterizan por diferentes modos de innovación según la presencia o ausencia de algunas condiciones de contexto que permitan la creación o adopción del conocimiento y la innovación. Dadas las diferencias en la intensidad del conocimiento y la innovación que caracteriza a cada modo/patrón de innovación, se formulan expectativas sobre la consecución del aumento de la productividad en las regiones con modelos de innovación diferentes. Los datos empíricos respaldan las expectativas conceptuales y muestran que en los patrones regionales de innovación basados en procesos de creación de conocimiento científico de ámbito local, existen rendimientos positivos al conocimiento científico. Sin embargo, en los patrones con menos intensidad de conocimientos también se pueden lograr pingües aumentos de productividad a través de actores locales que se aprovechan de conocimientos no científicos y capacidad innovadora. Es importante señalar que mientras que los beneficios derivados del conocimiento parecen ser bastante selectivos y concentrados en un número relativamente pequeño de regiones, los beneficios generados por la innovación parecen ser más dominantes y beneficiosos, incluso en regiones con un nivel bajo de conocimiento científico. Estos resultados tienen implicaciones importantes para el actual diseño de las políticas de innovación de la Unión Europea de ámbito regional.

INTRODUCTION

The importance of knowledge and innovation in the explanation of regional economic performance, structural disparities and productivity gap has progressively gained attention in both the scientific and the policy domains (EASTERLY and LEVINE, 2001; CAPELLO *et al.*, 2010, COOKE *et al.*, 2011; DETTORI *et al.*, 2011; PACI and MARROCU, 2013).

The Lisbon Agenda (EUROPEAN COUNCIL, 2000) and the Europe 2020 strategy (COMMISSION OF THE EUROPEAN COMMUNITIES, 2010a) goals of making Europe and its regions the most dynamic and competitive world economy move precisely in this direction by putting investments in knowledge creation at the core of a smart growth strategy. However, progress towards the achievement of the ambitious goals put forward in these key programmatic policy documents seems quite modest overall and, from a spatial perspective, highly selective and unbalanced (CAPELLO and LENZI, 2013a). Therefore, the occasion for relaunching a knowledge-intensive growth model for Europe on a regional base, opened by the new process of European Union Regional Policy Reform initiated in preparation of the new programming period 2014–20, calls for additional investigations on how research and innovation policies can be adapted to a regional setting, i.e., to a regional domain – defined by the smart specialization as the context or the setting in which entrepreneurship operates (FORAY *et al.* 2009; COMMISSION OF THE EUROPEAN COMMUNITIES, 2012; MCCANN and ORTEGA-ARGILÉS, 2014).

Targeted innovation policies are therefore required, and for this reason in-depth analyses on the efficiency of knowledge and innovation in different regional innovative contexts are necessary. The present paper aims to make a contribution in this regard by analysing, at the regional level, the impact of knowledge and innovation on the efficiency of local economic systems. By building on the quite extensive literature on the role played by knowledge and innovation on regional economic performance, the paper specifically looks for spatial heterogeneity in the role that knowledge and innovation may have in different territorial innovative contexts. In fact, regions can be characterized by different intensities in knowledge and innovation, and by different preconditions supporting the creation and adoption of knowledge and innovation. The way in which knowledge and innovation merge together and are supported and stimulated by specific territorial preconditions generates what we call a territorial pattern of innovation, i.e., a mode of innovation (CAPELLO, 2013).

In particular, this paper seeks to determine whether and to what extent regions characterized by different patterns of innovation are able successfully to exploit knowledge and innovation to achieve higher productivity levels. The conceptual expectation is that whereas the productivity accruing from knowledge are highly selective in space and concentrated in a handful of regions already showing a larger knowledge endowment, the benefits coming from innovation are spatially more diffusive and reach a larger group of regions, regardless of their knowledge and innovation intensity. This expectation comes from our belief that innovation does not necessarily require scientific and formal knowledge: there are patterns of innovation based on either non-scientific, informal knowledge or/and knowledge spillovers that might be as efficient in exploiting innovation for productivity gains than the modes of innovation based on scientific internal knowledge. For this reason, the distinctions between scientific versus non-scientific knowledge, on the one hand, and between knowledge and innovation, on the other, drive the analysis.

In so doing, this paper adds to existing studies in two main directions. On the scientific terrain, it departs from most traditional approaches assuming undifferentiated (positive) impacts of knowledge and innovation on productivity across regions. Moreover, it enriches the existing attempts in the literature that try to capture the spatial heterogeneity of the impact of knowledge and innovation on economic productivity by avoiding regional differentiations lacking robust and direct links to specific characteristics of each regional innovation mode (STERLACCHINI, 2008; CHARLOT *et al.*, 2012; MARROCU *et al.*, 2013). In this study, spatial heterogeneity is captured by differentiating regions according to their modes of innovation, which mostly depend upon specific and systemic context conditions for knowledge creation and for turning knowledge into innovation. This is in line with the branch of literature that emphasizes that 'regionalization is at least part of the solution to understanding dynamic industrial development in some places' (ASHEIM and ISAKSEN, 2002, p. 77). With respect to that branch of literature, our approach insists on the separation between the knowledge and the innovation phase, and enters in depth in the exploration of which local resources and regional assets represent the preconditions for the different phases of the innovation process to occur; the presence/absence of these preconditions support specific firms' innovative behaviours and approaches, giving rise to specific innovation modes at the regional level. The results show that remarkable differences exist in the ways in which firms belonging to specific regional innovation patterns are able to take advantage of existing resources such as knowledge and innovation. Although it is acknowledged that economic agents within each region have a large variety of innovative behaviours (SRHOLEC and VERSPAGEN, 2012), the ultimate aim of this paper is to direct innovation strategies at the regional level, and thus the interest lies in getting results on average regional innovative behaviours.

The need to orient innovation strategies also explains the choice of the NUTS-2 level of analysis, an administrative unit to which smart specialization strategies (and European regional policy more generally) are addressed in most European countries. In fact, on the policy terrain, the findings are of primary importance for the current efforts to design research and innovation policies within the frame of the Europe 2020 strategy to make the European Union and its regions grow smartly (COMMISSION OF THE EUROPEAN COMMUNITIES, 2010a). The differentiated impact of (different types of) knowledge and innovation across the European territory suggests and orients research and innovation strategies, targeted on different types of regions, consistently with the proposals developed in the smart specialization debate (FORAY *et al.*, 2009; CAMAGNI and CAPELLO, 2013; MCCANN and ORTEGA-ARGILÉS, 2014).

The reminder of the paper is organized as follows. The second section comments and elaborates on the different modes of regional innovation and their different expected effects on regional productivity. It introduces the concept of regional patterns of innovation and, accordingly, formulates the research hypotheses. The third section presents the territorial patterns of innovation taxonomy in European regions. The fourth section discusses the elaboration of the productivity measures used in the paper, i.e., the estimation of total factor productivity (TFP). The fifth section sets out the empirical framework. The sixth section comments on the empirical findings. The seventh section concludes and outlines some policy implications deriving from the results.

TERRITORIAL PATTERNS OF INNOVATION AND FACTOR PRODUCTIVITY: TESTABLE HYPOTHESES

Whereas most previous studies on the impact of knowledge and innovation on regional performance have generally focused on the effect of knowledge and innovation for the average region, very recently some scholars have tried to determine the possible differentiated impact across regions of these factors (e.g., STERLACCHINI, 2008; CHARLOT *et al.*, 2012; MARROCU *et al.*, 2013), suggesting that accounting for spatial heterogeneity is important to gain full understanding of the differentiated pathways from knowledge and innovation to higher inventive and economic performance.

Yet spatial heterogeneity has been captured with rather basic distinctions among regions. One criterion has been the socio-economic level of regions, which divides the European territory between advanced and disadvantaged regions defined as Objective 1 regions through political decisions (STERLACCHINI, 2008; CHARLOT *et al.*, 2012);[1] another criterion has been based on geographical macro-areas (institutional) boundaries, namely the distinction between EU-15 and EU-12 (MARROCU *et al.*, 2013; PACI and MARROCU, 2013). These definitions, however, lack robust links to specific characteristics of each region, and they are therefore unlikely to capture the related sources of regional differentiation.

In particular, these socio-economic and institutional distinctions cannot be used to capture important heterogeneity sources across regions in the types of knowledge needed for innovating and in the propensity to innovate, which mostly depend upon specific and systemic context conditions and tend to be highly cumulative and persistent over time (IAMMARINO, 2005).

The conceptual framework of analysis moves in this direction, based on the strong belief that regions innovate through different modes, according to the presence/absence of some context conditions that allow for the creation and/or adoption of knowledge and innovation. In fact, it is possible to consider alternative situations where innovation may build on an internal knowledge base, or on local creativity even in absence of local knowledge, or on innovative applications of a knowledge developed elsewhere and acquired via scientific linkages, or finally on imitative processes.

The conceptual framework of analysis is based on the notion of territorial patterns of innovation, defined as a combination of *context conditions* and of *specific modes of performing the different phases* of the innovation process (CAPELLO, 2013). The different phases of the innovation process refer to a logical sequence between knowledge, innovation and economic performance, as in the abstract but consistent 'linear model of innovation' – even if heavily criticized as unrealistic and rooted in the idea of a rational and orderly innovation process (EDGERTON, 2004). In fact, the authors strongly believe the following:

- In many cases scientific advance is a major source of innovation, as the information and communication technology paradigm and trajectory indicate.
- An alternative model of full complexity, where 'everything depends on everything else', does not help in conceptualizing and interpreting the systemic, dynamic and interactive nature of innovation.
- Self-reinforcing feedbacks from innovation to knowledge and from economic growth to innovation and knowledge play an important role in innovation processes.

The impact of science on innovation does not merely reside in the creation of new opportunities to be exploited by firms, but rather in increasing productivity of, and returns to, research and development (R&D) through the solution of technical problems, elimination of research directions that have proven wrong and the provision of new research technologies (NELSON, 1959; MOWERY and ROSENBERG, 1998; BALCONI *et al.*, 2010). We therefore strongly support the concept of a 'spatially diversified, phase-linear, multiple-solution model of innovation', in which the single patterns represent a linearization, or a partial block-linearization, of an innovation process where feedbacks, spatial interconnections and non-linearities play a prominent role.

The territorial innovation patterns concept stresses complex interplays between phases of the innovation process and the territorial context. In fact, each territorial pattern of innovation is a combination of different phases of the linear model of innovation and the presence/absence of territorial preconditions allowing for a certain phase to take place.

The concept of territorial patterns of innovation is in line with the idea developed in the literature that specific local and regional resources are important in firms' efforts to innovate, and therefore in forging firms' innovative behaviours (PIKE and TOMANEY, 1999; ASHEIM and ISAKSEN, 2002). In the regional innovation system (RIS) approach high proximity between different actors makes it possible for them to create, acquire, accumulate and utilize knowledge in faster ways than firms located outside knowledge-intensive, dynamic regional clusters. This definition of RIS explains why RIS is not a useful theoretical construct to study industrial development strategies in peripheral areas and declining industrial regions dominated by branch plant activities of transnational corporations (TNCs) (ASHEIM and ISAKSEN, 2002). The concept of territorial patterns of innovation differs from RIS in that it goes more in depth in the identification of the local elements that help firms to innovate, by distinguishing between territorial and local preconditions explaining firms' capacity to create knowledge, to turn knowledge into innovation, to acquire knowledge and innovation from outside the region. By doing so, the typology of innovative regions proposed is based on the presence/absence of territorial conditions that support specific firms' behaviours and propensity to innovate with respect to others. All types of innovative modes are taken into consideration, even the imitative innovation approaches typical of peripheral areas and declining industrial regions, frequently dominated by branch plant activities of TNCs (ASHEIM and ISAKSEN, 2002).

The approach strongly relies on the conceptual distinction between knowledge and innovation: they are conceptualized as different (and subsequent) logical phases of an innovation process, each phase requiring specific local elements for its development. This approach refuses the generalization of an invention–innovation short-circuit taking place inside individual firms (or territories), as that visible in some advanced sectors, as well as the assumption of an immediate interaction between R&D/high education facilities, on the one hand, and innovating firms, on the other, because of pure spatial proximity.

In this way, it is possible to separate the different necessary context conditions that support the single innovation phases and that generate *different modes of performing and linking-up the different phases of the innovation process*. These context conditions become integral parts of each territorial pattern of innovation.

A micro-founded approach drives the conceptualization of the territorial patterns of innovation; certain territorial resources and conditions influence certain types of firms' strategies and behaviours. High-technology firms, whose R&D activities rely on formal scientific and technological knowledge, look for local context conditions that support their internal R&D activities; a vibrant innovative environment, characterized by the presence of universities, R&D laboratories and an advanced scientific labour market, supports the development of codified scientific and technological knowledge. The high 'absorptive capacity' (à la COHEN and LEVINTHAL, 1990) in the area supports intense interregional cooperation behaviours for firms in search for complementary knowledge. The presence of such territorial elements give rise to an innovation mode similar to a science, technology and innovation (STI) mode of innovation (JENSEN *et al.*, 2007).

Creative small and medium firms, belonging to traditional rejuvenating sectors, in which tacit and often localized knowledge feeds the creativity of entrepreneurs is at the basis of another innovation mode; in such areas the creativity of local entrepreneurs pushes them to look for external knowledge and to use it for generating local innovation tailored on the firms' needs. This mode of innovation departs from the DUI model (JENSEN *et al.*, 2007) in that it considers the production of innovation based on both scientific and technological knowledge, as well as informal and tacit knowledge, produced *outside* the region, that is brought into the region, revisited, exploited and turned into innovation for internal needs thanks to the creativity of local firms.

A completely different firms' typology is behind an imitative innovation process: in contexts lacking endogenous conditions for knowledge and innovation creation, the only channels through which firms can innovate is through imitation. This is most likely to take place in traditional, stagnating sectors, or in branches of multinational firms, specialized in low-value functions.

On the basis of these possible typologies of firms' innovative behaviours, three types of innovation modes may be indicated (CAPELLO, 2013):

- *An endogenous innovation pattern in a scientific network*. In this pattern, highly innovative firms, belonging to high added value and technology-intensive sectors, are expected to cluster since they seek and require those local conditions – like the presence of universities, research centres, highly advanced human capital – that fully support the creation of knowledge. Moreover, in this group of regions the preconditions to turn knowledge into innovation, like the presence of entrepreneurial spirit and creativity, guarantee the transformation of knowledge into innovation. Given the complex nature of knowledge creation nowadays, this pattern is expected to show a tight interplay among regions in the form of international scientific networks. From the conceptual point of view, this advanced pattern is the one considered by most of the existing literature dealing with knowledge and innovation creation and diffusion. In these regions there is a higher presence of innovative firms with a strong capacity to absorb and use 'global' knowledge, similarly to the STI mode of innovation described by JENSEN *et al.* (2007).
- *A creative application pattern*, characterized by the presence of creative (small and medium) enterprises belonging to traditional or medium-technology sectors, curious enough to look for knowledge outside the region – given the scarcity of local knowledge – and creative enough to apply external knowledge to local innovation needs (FORAY, 2009; COMMISSION OF THE EUROPEAN COMMUNITIES, 2010b). Knowledge providers, supporting local firms' innovative activities, are mostly located outside the region and knowledge exchanges are mainly nourished by cognitive and sectoral proximity (i.e., shared cognitive maps) than by belonging to the same local community, as in the regionalized national innovation system type of RIS proposed by ASHEIM and ISAKSEN (2002).
- *An imitative innovation pattern*, where firms in traditional sectors, or branches of multinational enterprises, in different sectors, seeking for low labour costs areas to locate their lower value-added functions, base their innovation capacity on imitative processes, which can take place with different degrees of adaptation on already existing innovations. This pattern is based on the literature dealing with innovation diffusion. In several cases, regions in this pattern are likely to be characterized by a higher presence of firms with little learning and innovative activities, similarly to the low learning innovative mode described by JENSEN *et al.* (2007).

When applied at the empirical level, therefore, this framework of territorial innovation patterns can provide a new and interesting lens through which one can assess productivity gains that come from the specific combination of context conditions and modes of performing knowledge and innovation processes that each region may exhibit. In particular, given the differences in the intensity of knowledge and innovation, the following may be expected:

- Regions characterized by a higher intensity of scientific knowledge creation achieve higher productivity gains from this knowledge than less knowledge-intensive regions, consistent with the absorptive capacity argument proposed by COHEN and LEVINTHAL (1990). In fact, *increasing returns are expected to exist in knowledge exploitation when universities, research centres and highly qualified human capital are available in a region.*
- Regions able to innovate through non-scientific knowledge or through knowledge spillovers obtained from other regions achieve productivity gains from both non-scientific knowledge and innovation. In fact, it is expected that *innovation not only derives from scientific knowledge and inventions but also from non-scientific knowledge or from knowledge spillovers obtained from other regions, when creativity and entrepreneurship are present in an area.*
- No a-priori exists for the effects of innovation on the economic performance of imitative regions. In fact, *the efficiency of imitation is expected to rest on the degree of innovativeness of imitation processes.*

Hence, from this conceptual framework, the following testable hypotheses can be formulated:

Hypothesis 1: Regions with higher knowledge and innovation intensity can achieve higher productivity levels; however, regions innovating in the absence of a strong scientific knowledge base can achieve productivity gains comparable with more knowledge-intensive innovating regions.

Hypothesis 2: Scientific knowledge has a greater impact on economic performance of knowledge-intensive regions (i.e., belonging to the first conceptual pattern) than on the performance of non-scientific knowledge-intensive regions (i.e., belonging to the second and third conceptual patterns).

Hypothesis 3: Non-scientific knowledge has a greater impact on the economic performance of regions belonging to the second and third conceptual patterns than on knowledge-intensive regions (i.e., belonging to the first conceptual pattern).

Hypothesis 4: Innovation has an impact on the economic performance of all regions and not only of knowledge-intensive regions.

The economic performance measure is represented by total factor productivity (TFP), for two major reasons. Firstly, there is no a-priori on whether innovation processes analysed are capital or labour saving, and for this reason measures like labour productivity only would be not fully appropriate or even misleading. Secondly, efficiency, rather than effectiveness, is more appropriate when innovation processes are analysed. In fact, while efficiency can be certainly associated with innovation processes, effectiveness is not mechanically the result of an innovation process. An example is represented by Eastern European regions that, before the crisis, were registering the highest economic growth rate in Europe without deep innovation processes.

TFP, therefore, seems to fit at best our goal. The next section deals with the identification of territorial patterns of innovation in Europe, while the fourth and fifth sections present the TFP estimation and the empirical model respectively.

TERRITORIAL PATTERNS OF INNOVATION IN EUROPE

At the empirical level, the territorial patterns of the innovation framework suggest a partitioning of the European space that shows some advantages with respect to other empirical research works. Most of the existing taxonomies, in fact, group European regions only on the basis of their intensities of knowledge production, taking it for granted that knowledge leads (and equates) to innovation while ignoring the territorial conditions behind local innovation modes. This is precisely the case of the most recent Organisation for Economic Co-operation and Development (OECD) taxonomy (OECD, 2010, 2011), which identifies 'knowledge regions, industrial production zones, non-S&T [science and technology] driven regions' on the basis of R&D intensity. Although important and interesting results are achieved, the methodologies employed merge together indicators as diverse as innovation performance, knowledge inputs like R&D, sectoral structure and the presence of spatial innovation enablers, with no clear conceptual expectations on the linkages among the different variables, in a purely inductive way (HOLLANDERS *et al.*, 2009a; WINTJES and HOLLANDERS, 2010). In the end, many of these partitioning exercises are mostly data driven rather than theoretically driven. The own goal, on the contrary, is to detect regional 'patterns' based on a clear conceptual definition of the different phases of any innovation process, and of the context conditions that are expected to support the different phases of the innovation process.

In a recent work, territorial patterns of innovation have been empirically detected by means of a *k*-means cluster analysis based on a series of indicators capturing the different knowledge and innovation propensity across European regions: namely the regional European Union share of total patents, the regional share of firms introducing product and/or process innovation and the regional share of firms introducing marketing and/or organizational innovation (CAPELLO and LENZI, 2013b).[2] The empirical results show a larger variety of possible innovation patterns than the ones conceptually envisaged, still consistent with the theoretical underpinnings presented before. Two clusters can be associated with the first conceptual pattern, albeit with some relevant distinctions between the two; two clusters can be associated with the second pattern, again with some important differences; and one cluster can be associated with the third pattern. In particular:

- A *European science-based area (ESBA)*, composed of strong knowledge and innovation-producing regions, specialized in general-purpose technologies, with the highest generality and originality of the scientific local knowledge, and the highest degree of knowledge acquisition from other regions. R&D expenditures, too, are the highest in these regions. They are mostly located in Germany, with the addition of Vienna (Austria), Brussels (Belgium) and Syddanmark (Denmark).
- An *applied science area (ASA)*, similarly made up of strong knowledge-producing regions, albeit characterized by a local knowledge base of an applied nature, and by a high degree of knowledge acquisition from other regions. R&D activity is high in this group of regions as well. Regions of this type are mostly agglomerated and located in Central and Northern Europe, namely in Austria, Belgium, Luxembourg, France (i.e., Paris), Germany, Ireland (i.e., Dublin), Denmark, Finland and Sweden, with some notable exceptions to the East such as Prague (Czech Republic), Cyprus and Estonia and to South such as Lisbon (Portugal) and Attiki (Greece).
- A *smart technological application area (STAA)*, with a high product innovation rate, with a limited degree of local basic science, but a high level of creativity, with respect to the other four clusters, which enables the translation of external basic and applied scientific knowledge into innovation. The knowledge intensity is lower than in the previous two cases, although not negligible. This group of regions includes mostly agglomerated regions in the EU-15, such as the northern part of Spain and Madrid, Northern Italy, the French Alpine regions, the Netherlands, Czech Republic, Sweden and the UK.
- A *smart and creative diversification area (SCDA)*, with a low degree of local scientific knowledge in the form of patents and R&D, a non-negligible internal innovation capacity, a high degree of local capabilities (i.e., non-scientific and tacit knowledge embedded in professionals), of creativity and entrepreneurship, and of acquisition of external knowledge embedded in professional capabilities, with respect to the other four clusters. These regions are mainly located in Mediterranean countries (i.e., most of the Spanish regions, Central Italy, Greece, Portugal), in EU-12 agglomerated regions in Slovakia and Slovenia, Poland and Czech Republic, and in a few regions in Northern Europe, namely in Finland and the UK.
- An *imitative innovation area (IIA)*, characterized by a low knowledge and innovation intensity but high entrepreneurship, creativity, attractiveness and innovation potentials. Most of these regions are in the EU-12, such as all regions in Bulgaria and Hungary, Latvia, Malta, several regions in Poland, Romania and Slovakia, and also in Southern Italy.

The five groups therefore differ not only in terms of their knowledge and innovation endowments, but also, interestingly, in the type and nature of the knowledge used in innovative activities and in the enabling territorial factors supporting the creation and acquisition of different types of knowledge and its successful conversion into innovation. Whereas scientific and technical knowledge (being either generic or specific), as captured by patents and R&D expenditures, is prominent in the first three groups, in the last two the relevant knowledge is less science and technology based, i.e., embedded in the human capital of specialized workers.

TOTAL FACTOR PRODUCTIVITY (TFP) ESTIMATION

Estimation of TFP has been a long-debated and highly controversial topic in economics. Three main approaches can be identified in the literature and, to date, no definitive consensus on a single method has been reached, since all approaches have shortcomings and are subject to criticisms (for a survey, see CARLAW and LIPSEY, 2003).

Initially, the most common method of calculating TFP was growth accounting (or residual method) based on SOLOW's (1956, 1957) seminal papers. In this approach, TFP is 'a measure of our ignorance', and therefore obtained as the residuals of output regressed on inputs. This methodological approach has been heavily criticized in the literature in terms of both concepts (e.g., the aggregation of the production function) and measurement (e.g., measurement of inputs and output) (e.g., GRILICHES 1987, 1995), although it is still well maintained and largely used (OECD, 2001; QUATRARO, 2010; DETTORI *et al.*, 2011; MARROCU *et al.*, 2013).

A different approach is the index number approach in which TFP is computed as the ratio between output and an index aggregating all production factors (DIEWERT, 1987). This approach overcomes the conceptual ambiguity of the aggregate production function inherent to the growth-accounting approach, at the cost, however, of identifying the appropriate index adding up all inputs into a single scalar.

Lastly, the distance function approach is based on the measurement of the distance from the actual production of an economy to the production efficiency frontier. For example, the data envelopment analysis (DEA) approach, first developed by FARRELL (1957), belongs to this group. This is a non-parametric method aimed at identifying the best performing units whose objective is to convert multiple inputs into multiple outputs. DEA

Table 1. Total factor productivity (TFP) estimation: data description

Indicators	Measures	Computation	Year	Source
Gross domestic product (GDP)	Economic wealth	€ (thousands)	Average, 2005–07	EUROSTAT
Capital	Capital stock (factor endowment)	€ (thousands)	Average, 2005–07	CRENoS database[a]
Labour	Employment (factor endowment)	Unit of labour (thousands)	Average, 2005–07	EUROSTAT

Notes: [a]The authors are grateful to the Centro Ricerche Economiche Nord Sud (CRENoS), University of Cagliari, Italy, for granting access and use of its capital series data.

EUROSTAT, Statistical Office of the European Communities.

Table 2. Total factor productivity (TFP) estimation: descriptive statistics

Variable	Number of observations	Mean	Standard deviation (SD)	Minimum	Maximum
Gross domestic product (GDP)	259	43682.15	50 926.75	1070.608	500 369.8
Labour	259	822.10	655.33	14	5122.23
Capital	259	9009.29	11 437.51	92	110 349.7

does not require specification of the functional form for the relation linking inputs to outputs, and it is able to handle multiple inputs and outputs. The best performance is characterized in terms of efficiency, so that the most performing units define the efficient frontier, which 'envelopes' all the other units. Nevertheless, this approach critically relies on the assumption that all units share the same aggregate production function linking inputs and outputs.

It is therefore evident that no method is free from limitations. In the analysis reported in this paper, the standard method was adopted, i.e., a traditional growth-accounting approach to capital–labour multifactor productivity (OECD, 2001). As discussed in the second section, TFP was chosen as a measure of economic performance since there is no a-priori on whether innovation processes analysed are capital or labour saving, and for this reasons measures like labour productivity only would be not fully appropriate or even misleading. The regional TFP level, thus, was estimated as the residual of the log-linearized version of the traditional Cobb–Douglas production function model, taking the following form:

$$GDP_r = \alpha + \beta K_r + \gamma L_r + \varepsilon_r \qquad (1)$$

where GDP_r is the regional gross domestic product; K_r is the regional capital stock; L_r is the regional employment level; and ε_r represents the regional idiosyncratic error. GDP, capital stock and labour were averaged over the years 2005–07 to smooth possible effects related to specific years of estimation. The capital stock series at the regional level is not available from public databases and official sources. The capital stock series was elaborated by the Centro Ricerche Economiche Nord Sud (CRENoS), University of Cagliari, Italy, and was constructed by applying the perpetual inventory method on investment series in the years 1985–2007. Specifically, K_r, the capital stock of region r at time t, is obtained as the sum of the flows of gross investments in the previous periods with a constant (across regions and over time) 10% depreciation rate (d), as is customary in this kind of exercise (MARROCU and PACI, 2011), as follows:

$$K_{r,t} = (1 - d)K_{r,t-1} + I_{r,t-1} \qquad (2)$$

The capital stock value for the initial year (i.e., 1985) was computed as the sum of investment flows, $I_{r,t}$, in the ten preceding years (i.e., 1975–1984). Tables 1 and 2 report the inputs and output variables description and summary statistics respectively.

Estimates were obtained by ordinary least squares (OLS), spatial error model (SEM) and spatial lag model (SAR). Given the significance of robust Lagrange multipliers for both SEM and SAR, SEM was preferred to SAR because of the larger statistic value, suggesting that probably the spatial error model is more adequate than the spatial lag model. Still, TFP estimation obtained through different methods is highly consistent being the correlation among them greater than 0.93. Factor endowments' coefficient estimates are reported in Table 3.

THE DETERMINANTS OF TFP AND THE EMPIRICAL MODEL

The empirical investigation of the research hypotheses advanced in the second section has quite strong requirements in terms of data availability and calls for a rich dataset that makes it possible to estimate productivity and measure several groups of factors that may affect it.

In particular, the role of knowledge and innovation was focused on. The impact of knowledge on productivity increase (captured mostly through either patent-related indicators or R&D measures) has been consistently detected in the empirical literature (MADSEN, 2008; QUATRARO, 2010). Accordingly, a

Table 3. Total factor productivity (TFP) estimation: estimated coefficients

	OLS	SEM	SAR
Capital (average, 2005–07)	0.594***	0.342***	0.352***
	(0.052)	(0.040)	(0.040)
Labour (average, 2005–07)	0.339***	0.609***	0.565***
	(0.066)	(0.50)	(0.50)
Constant	2.945***	3.335***	−2.972***
	(0.324)	(0.622)	(0.622)
Lambda (SEM)/Rho (SAR)		0.962***	0.639***
		(0.031)	(0.054)
R^2 (OLS) – squared correlation (SEM and SAR)	0.75	0.72	0.72
Number of observations	262	262	262
Robust Lagrange multiplier (error)	424.84***		
Robust Lagrange multiplier (lag)	17.372***		

Note: $*p < 0.10$, $**p < 0.05$, $***p < 0.01$. SEM and SAR estimates are based on a row-standardized continuous-distance matrix.

positive impact of the knowledge variables on productivity was expected. As a novelty of the conceptual approach and empirical strategy, however, a distinction between scientific knowledge and non-scientific knowledge was introduced, the latter embedded in specific professional capabilities. Moreover, thanks to the availability of a rich dataset on innovation, the knowledge and innovation variables were introduced separately, the expectation being different spatial effects for these variables.

Summing up, as regards knowledge and innovation, three variables were considered that capture, respectively, the intensity of:

- scientific knowledge, measured through patent intensity (i.e., the count of European Patent Office (EPO) patent applications on the population);
- non-scientific knowledge embedded in capabilities measured through an indicator derived from factor analysis aimed at capturing the knowledge embedded in the human capital of specialized workers such as managers and technicians (i.e., professional capabilities); and
- innovation, to account directly for the impact of new products and/or processes introduced in the market (i.e., the share of firms introducing product and/or process innovation).

When measuring the impact of knowledge and innovation on TFP level across the five territorial patterns of innovation, however, we also controlled for other factors that recent studies have shown to be relevant in the process of knowledge creation and diffusion, and that are therefore important determinants of TFP level and growth in the current years, namely specialization externalities that operate mainly within a specific industry, agglomeration externalities that work across sectors, and social capital that operates across individuals (BEAUDRY and SCHIFFAUEROVA, 2009; DETTORI *et al.*, 2011; MARROCU and PACI, 2011; MARROCU *et al.*, 2013).

The most common way to measure regional industrial specialization is by means of a location quotient on employment in manufacturing sectors. The long-standing debate on the relative importance of specialization versus diversification externalities (BEAUDRY and SCHIFFAUEROVA, 2009) indicates that empirical evidence is not robust in this regard and that results are overall ambiguous. Some studies report positive effects of specialization (e.g., HENDERSON *et al.*, 2001; CINGANO and SCHIVARDI, 2004), whereas others report negative effects of specialization (FRENKEN *et al.*, 2007; PACI and USAI, 2008). Although, at a conceptual level, the expected impact of specialization is rather unpredictable in that it depends on the level of analysis, unit of measurement and data used (BEAUDRY and SCHIFFAUEROVA, 2009), the development of the service sector affecting most Western European countries and the prominent localization of manufacturing activities in new EU-12 countries suggest that this variable has a negative effect on TFP level (MARROCU *et al.*, 2013).

To capture the synergies, complementarities, collective learning effects and local knowledge spillovers arising in dense agglomerations of different economic activities, which are at the basis of knowledge and innovation creation, and possible productivity gains, a variable was introduced to account for population density in the region (labelled 'agglomeration economies'). This variable is measured as thousands of inhabitants per square kilometre in the regions. The empirical literature has consistently shown the importance of agglomeration economies for productivity increases and more generally for economic performance (BEAUDRY and SCHIFFAUEROVA, 2009). Therefore, this variable is expected to have a positive effect on TFP level.

If knowledge spillovers at sectoral level are important determinants of TFP, also the degree of cooperative propensity and collective actions within a region is a condition for local knowledge spillovers. Higher cooperative propensity should promote knowledge and innovation circulation and socialization, thus enhancing productivity. Similarly, trust – and therefore openness to diversity – characterizes settings that are more innovation prone and thus more rapid in adopting innovation and profiting from the productivity gains it yields. Trust, in fact, has been proved to facilitate cooperation (GUISO *et al.*, 2008), to reduce transaction costs and opportunistic behaviour, thereby promoting knowledge and innovation diffusion and an overall improvement in economic environment and performance. This variable is therefore expected to have a positive effect on TFP level. In the present case, the regional share of people trusting each other was the indicator used to measure social capital.

Furthermore, a control for the distinction between EU-12 and EU-15 was introduced in our model, because TFP level and dynamics have been shown to differ quite sharply according to the development stage of economies; in particular, productivity levels tend to be higher in more developed areas, despite the rapid catch-up registered in EU-12 (MARROCU *et al.*, 2013). Country dummies were also added to control for national variations that may affect the nature and institutional setting of regional innovation processes, the general economic dynamism, as well as the nature and pattern of development, and possible economic interdependencies across regions in the same country.

Lastly, a set of dummy variables accounting for the territorial patterns of innovation described in the third section was included in order to capture the impact of different knowledge and innovation intensities across regions. Accordingly, the following baseline model was estimated:

$$\text{TFP}_r = F(\text{Specialisation in manufacturing}_r, \text{Agglom}, \text{Trust}_r, \text{EU12}, \text{SCDA}_r, \text{STAA}_r, \text{ASA}_r, \text{ESBA}_r) + \varepsilon_r \quad (3)$$

where SCDA_r, STAA_r, ASA_r and ESBA_r are the dummies each capturing the impact of the different territorial patterns of innovation with respect to the reference category (i.e., the *Imitative innovation area*, IIA_r).

To unravel further the spatial heterogeneity in productivity gains across patterns, the baseline model was modified to take into account the different intensity of regional knowledge and innovation. The five dummies conceptually differentiate the impact for the different modes of innovation; practically, they allow one to differentiate the impact of each variable at different degrees (i.e., ranks) of knowledge and innovation intensity. Technically speaking, in equation (3) the territorial innovation patterns dummies were interacted with the scientific and non-scientific knowledge and innovation variables, in turn, as follows:[3]

$$\text{TFP}_r = F(\text{Specialisation in manufacturing}_r, \text{Trust}_r, \text{EU12}, \text{Agglom}_r, \text{PAT}_r, \text{PAT}*\text{SCDA}_r, \text{PAT}*\text{STAA}_r, \text{PAT}*\text{ASA}_r, \text{PAT}*\text{ESBA}_r) + \varepsilon_r \quad (4)$$

$$\text{TFP}_r = F(\text{Specialisation in manufacturing}_r, \text{Trust}_r, \text{EU12}, \text{Agglom}_r, \text{CAP}_r, \text{CAP}*\text{SCDA}_r, \text{CAP}*\text{STAA}_r, \text{CAP}*\text{ASA}_r, \text{CAP}*\text{ESBA}_r) + \varepsilon_r \quad (5)$$

$$\text{TFP}_r = F(\text{Specialisation in manufacturing}_r, \text{Trust}_r, \text{EU12}, \text{Agglom}_r, \text{INNO}_r, \text{INNO}*\text{SCDA}_r, \text{INNO}*\text{STAA}_r, \text{INNO}*\text{ASA}_r, \text{INNO}*\text{ESBA}_r) + \varepsilon_r \quad (6)$$

where PAT_r, CAP_r and INNO_r are the regional patent intensity, professional capability and innovation indicators, respectively. Four interaction terms are inserted, to be interpreted with respect to the reference category (i.e., the *IIA*). These three models were estimated in turn to avoid possible multicollinearity issues.

Tables 4 and 5 report the description of the variables and their sources and their summary statistics respectively.

TFP DETERMINANTS ACROSS TERRITORIAL PATTERNS OF INNOVATION

Table 6 reports the estimates of TFP level determinants.[4] The control variables are significant and with the expected sign. Consistently with previous studies (among many others, see MARROCU *et al.*, 2013) TFP is higher in the EU-15; it benefits from agglomeration economies and a higher level of trust, whereas it is negatively affected by specialization in manufacturing. As discussed above, the empirical evidence on this specific result is not always convergent, and the literature provides several examples of both positive and negative impacts of specialization on economic performance (BEAUDRY and SCHIFFAUEROVA, 2009).

More interestingly, the dummies representing the different territorial patterns of innovation show a positive and significant sign with respect to the reference category, i.e., the *IIA* (models 1 and 2 in Table 6). Although the five groups are characterized by different (and increasing) levels of knowledge and innovation, each innovation mode can lead to productivity increases, as expected by Hypothesis 1. This result suggests that the productivity level of European regions is not linked only to the strength of the local scientific knowledge base.

Interesting results emerge on inspecting the results of TFP in the different patterns of innovation more closely

Table 4. Description of variables

Indicators	Measures	Computation	Year	Source
Total factor productivity (TFP)	Economic productivity	Residuals	Average, 2005–07	Authors' estimation on EUROSTAT data
Specialization in manufacturing	Location quotient in manufacturing sectors computed on employment data	Location quotient on employment in manufacturing sectors	2002	EUROSTAT
EU-12	Bulgaria, Cyprus, Czech Republic, Hungary, Estonia, Latvia, Lithuania, Malta, Poland, Romania, Slovakia, Slovenia	Dummy variable equal to 1 if the regions is located in an EU-12 country	2004	EUROSTAT
Social capital	Trust	Share of people trusting each other	2000	European Value Survey
Agglomeration economies	Population density	Thousands of inhabitants per km^2	1999–2001	EUROSTAT
Knowledge intensity	Patent intensity	Total patents on population	1999–2001	CRENoS database[a]
Capabilities	Share of managers and technicians	Factor analysis on the share of production and specialized service managers and science and engineering associate professionals (ISCO codes 13 and 31, respectively)	Average, 1997–2001	European Labour Force Survey[b]
Innovation (product and/or process)	Firms introducing a new product and/or a new process in the market	Share of firms introducing a product and/or process innovations	One value for 2002–04	Authors' estimation based on the CIS (EUROSTAT) database[c]

Notes: [a]Given the erratic nature of patent counts, the 1999–2001 patent stock was considered. Patents were assigned to regions according to the respective inventors' residence address as available in patent documents. Fractional count was applied. The authors are grateful to the Centro Ricerche Economiche Nord Sud (CRENoS), University of Cagliari, Italy, for granting access and use of its patent database.

[b]Variable capabilities summarize the share of production and specialized service managers and science and engineering associate professionals (International Standard Classification of Occupations (ISCO) codes 13 and 31, respectively). This variable is built around the definition of human resources in science and technology (HRST) in terms of occupation provided by the Statistical Office of the European Communities (EUROSTAT), which separates out scientific and technical occupations (considered as HRST) from managerial occupations (considered as not being part of HRST). The variable therefore tries to capture the divide between scientific and engineering (i.e., technological) competencies, on the one hand, and managerial competencies, on the other. This dichotomy aims at characterizing the nature of the regional knowledge base, i.e., whether it is based on scientific and technical capabilities (i.e., scientific knowledge) or rather on managerial capabilities (i.e., non-scientific knowledge). This variable is obtained by factor analysis where one factor has been extracted through the principal component method. The percentage of variance explained is 0.67. The rotated factor loadings (with Varimax rotation and Kaiser normalization) are 0.82 for managerial competencies and –0.82 for scientific and technical competencies.

[c]See CAPELLO and LENZI (2013a) for details on the estimation methodology. CIS is definitely an innovative action in many respects because, for the first time, it has made it possible to collect internationally comparable direct measures of innovative activities at a highly disaggregated level of analysis (FAGERBERG *et al.*, 2012). Despite these merits and the increasing worldwide development and use of similar innovation surveys based on the Oslo manual such as CIS (HONG *et al.*, 2012), CIS has been much criticized. In particular, at the regional level of analysis, CIS is affected by two specific problems. Firstly, because administration of the survey is implemented at the national level, and by national statistics offices, not all countries stratify the sample at the same regional level. Secondly, even if a regional sample stratification is adopted, non-representativeness bias may persist for small regions (HOLLANDERS *et al.*, 2009b). Notwithstanding the doubts concerning CIS data, the exercise presented here can be defended. In fact, despite the use of a different methodology, the results obtained are consistent with previous estimates implemented for the DG Industry and DG Regio (HOLLANDERS *et al.*, 2009a, 2009b; CAPELLO and LENZI, 2013a). The consistency captured by all robustness tests is a sufficiently strong scientific basis to claim that a proxy for innovation is available.

Table 5. Summary statistics

Variable	Number of observations	Mean	Standard deviation (SD)	Minimum	Maximum
Total factor productivity (TFP)	259	0.02	0.52	−1.68	1.37
Trust	259	0.31	0.16	0	0.82
Agglomeration economies	259	0.28	0.52	0.003	3.80
Specialization in manufacturing	259	0.993	0.329	0.302	1.930
Knowledge intensity	259	0.091	0.114	0	0.652
Capabilities	259	−0.058	0.933	−2.487	3.548
Innovation	259	35.57	13.29	7.97	87.10

by means of a pairwise *t*-test on the equality of coefficients. Firstly, the TFP efficiency is somewhat higher in more scientific knowledge-intensive regions. However, the efficiency ranking does not strictly reflect the knowledge intensity ranking in the form of R&D expenditures, thus supporting Hypothesis 1. In fact, despite relatively limited R&D efforts and patent intensity, the *SCDA* has a level of efficiency similar to that of the *STAA*. Likewise, the *ESBA* shows efficiency levels similar to those of the *ASA*, despite its far larger scientific knowledge endowment (Table 7).

Estimates of equations (3) to (5) are displayed in Table 6 in columns 3–6, respectively. Following ANSELIN (1988), Lagrange multiplier tests were used to check for possible spatial dependency in the estimates. The results of the tests (reported in the bottom lines of Table 6), however, do not support the use of spatial lag and spatial error models, except for model 4 which has been estimated also by SAR.

The first interesting result is that knowledge intensity has a positive and significant effect with respect to the *IIA*, but only in those groups of regions strongly endowed with scientific knowledge, namely the *ESBA*, the *ASA* and the *STAA*, where it substantially increases the TFP level (Table 6, column 3),[5] thus confirming Hypothesis 2. This result suggests that scientific knowledge is not the only and chief driver leading to higher productivity. Rather, a tight relationship between scientific knowledge and productivity level seems to operate only in those groups of regions in which the local knowledge base is already quite developed and rich. Scientific knowledge productivity gains are therefore highly selective from a spatial point of view. This finding is consistent with results obtained by STERLACCHINI (2008), CHARLOT *et al.* (2012), and CAPELLO and LENZI (2013c), suggesting that the impact of additional investments in R&D is greater in regions with a larger scientific knowledge base.

Interestingly, productivity gains can be achieved not only from further investments in scientific knowledge, but also from further investments in non-scientific and tacit knowledge embodied in human capital and professionals. In fact, the impact of capabilities on TFP is far lower with respect to the *IIA* (and becomes negligible, if not negative and significant) in the *STAA*, in the *ASA* and in the *ESBA*. Differently, knowledge productivity gains in the *SCDA* do not differ from those in the *IIA*, being thus linked to non-scientific and tacit knowledge embedded in professional capabilities rather than to scientific knowledge (Table 6, column 5), as expected by Hypothesis 3.

Lastly, innovation seems a more pervasive channel through which to achieve higher productivity levels because innovation has a significant and positive coefficient in all groups with respect to the *IIA* (Table 6, column 6), thus supporting Hypothesis 4.

On the one hand, productivity increases in *IIA* regions seem to respond at least (with respect to the other groups of regions) either to innovation or to knowledge; possibly these regions may not have yet reached a critical mass of both knowledge and innovation so as to turn their benefits into higher productivity levels. This is consistent with previous results in which innovation impact on GDP growth is characterized by similar threshold effects, since growth in the *IIA* is weakly responsive to increases in innovation (CAPELLO and LENZI, 2013c).

On the other hand, the impact of innovation on TFP is similar in the other four groups of regions, regardless of their knowledge and innovation intensity. Although the different sizes of the innovation coefficients estimates (Table 6, column 6) might suggest that innovation benefits are unevenly reaped by the different groups of regions, tests implemented on the significant coefficients of the innovation interaction variables indicate that, overall, these coefficients are not statistically different from one another, albeit they are all significantly different from the *IIA* group. Only one exception stands out, and it concerns the statistically significant difference between the *STAA* and the *ASA* coefficients ($F = 7.69$, $p < 0.01$), being the innovation impact smaller in the former than in the latter. All in all, therefore, innovation productivity gains are spatially more distributed, more pervasive and more even from a geographical point of view than are those stemming from knowledge, whether scientific (e.g., patent intensity) or non-scientific (e.g., capabilities).

Two important considerations can be drawn from these results. Firstly, the spatial distribution of the productivity gains arising from knowledge is selective and uneven and depends not only upon the strength of the local scientific knowledge (as developed through R&D activities and patents). Scientific knowledge is

Table 6. Determinants of total factor productivity (TFP) and territorial patterns of innovation

Dependent variable TFP, 2005–07	OLS (1)	SAR (2)	OLS (3)	OLS (4)	SAR (5)	OLS (6)
Specialization in manufacturing	–0.363***	–0.372***	–0.337***	–0.304***	–0.303***	–0.390***
	(0.063)	(0.054)	(0.062)	(0.063)	(0.052)	(0.067)
Trust	0.582***	0.585***	0.724***	0.769***	0.793***	0.603***
	(0.227)	(0.160)	(0.218)	(0.242)	(0.165)	(0.224)
EU-12	–0.380***	–0.427**	–0.121**	–0.134**	–0.146	–0.421***
	(0.070)	(0.210)	(0.051)	(0.060)	(0.204)	(0.065)
Agglomeration economies	0.078***	0.080***	0.118***	0.098***	0.100***	0.080***
	(0.032)	(0.028)	(0.032)	(0.037)	(0.028)	(0.030)
Smart and creative diversification area (SCDA)	0.266***	0.249***				
	(0.055)	(0.059)				
Smart technological application area (STAA)	0.313***	0.298***				
	(0.064)	(0.073)				
Applied science area (ASA)	0.477***	0.457***				
	(0.075)	(0.086)				
European science-based area (ESBA)	0.578***	0.559***				
	(0.115)	(0.108)				
Knowledge intensity			1.651			
			(2.308)			
Knowledge intensity*SCDA			0.701			
			(0.545)			
Knowledge intensity*STAA			0.613**			
			(0.310)			
Knowledge intensity*ASA			0.994***			
			(0.263)			
Knowledge intensity*ESBA			0.865***			
			(0.220)			
Capabilities				0.191**	0.207***	
				(0.079)	(0.082)	
Capabilities*SCDA				–0.126*	–0.126	
				(0.074)	(0.079)	
Capabilities*STAA				–0.201***	–0.199**	
				(0.074)	(0.084)	
Capabilities*ASA				–0.298***	–0.298***	
				(0.101)	(0.094)	
Capabilities*ESBA				–0.408***	–0.378***	
				(0.120)	(0.107)	
Innovation						–0.340
						(0.555)
Innovation*SCDA						0.995***
						(0.245)
Innovation*STAA						0.979***
						(0.299)
Innovation*ASA						1.280***
						(0.343)
Innovation*ESBA						1.205***
						(0.426)
Constant	0.171*	0.121	0.340***	0.363***	0.346***	0.268**
	(0.100)	(0.094)	(0.098)	(0.124)	(0.085)	(0.0110)
Country dummies	Yes	Yes	Yes	Yes	Yes	Yes
Rho (SAR)		0.210*			0.200**	
		(0.117)			(0.085)	
Test of joint significance of the coefficients of the variables Pattern 2, Pattern 3, Pattern 4 and Pattern 5 (models 1 and 2) and their interaction (models 3–6)	10.54***	30.91***	5.57***	4.56***	19.18***	5.98***
R^2 (OLS) – squared correlation (SAR)	0.86	0.86	0.86	0.85	0.86	0.86
Lagrange multiplier (spatial error)	0.233		0.148	1.597		1.046
Lagrange multiplier (spatial lag)	3.069*		0.386	6.043***		2.033
Number of observations	259	259	259	259	259	259

Notes: $*p < 0.10$, $**p < 0.05$, $***p < 0.01$. SAR estimates are based on a row-standardized continuous distance matrix.

ESBA, European science-based area; ASA, applied science area; STAA, smart technological application area; SCDA, smart and creative diversification area. The test of joint significance indicates that the dummy variables representing the territorial patterns of innovation are jointly statistically different from zero (models 1 and 2) and that the interaction variables between the dummy variables representing territorial patterns of innovation and, respectively, patent intensity (model 3), capabilities (models 4 and 5) and innovation (model 6) variables are jointly statistically different from zero.

Table 7. Pairwise comparison of impact on TFP level across territorial patterns of innovation

	ESBA	ASA	STAA	SCDA
European science-based area (ESBA)	–			
Applied science area (ASA)	n.s.	–		
Smart technological application area (STAA)	<***	<***	–	
Smart and creative diversification area (SCDA)	<***	<***	n.s.	–

Note: n.s., Not statistically significant, $^{*}p < 0.10$, $^{**}p < 0.05$, $^{***}p < 0.01$. Tests are implemented on model 2 of Table 6.

crucial but not exhaustive. In fact, higher productivity levels can be achieved also by taking advantage of local non-scientific, tacit and embodied knowledge creatively and successfully translated into commercially viable innovations. Whereas these results do not gainsay the importance of research activities for regional economic performance, and thus support the right focus put forward by the Lisbon Agenda and the Europe 2020 strategy on a 'smart growth' based on knowledge and the efforts to enlarge and strengthen the European knowledge base, they call for greater attention when this goal is translated into a regional setting. Different types of knowledge are in fact needed in different innovative contexts to improve productivity.

Secondly, innovation benefits are more pervasive than knowledge benefits because innovation is an important leverage of productivity gains across all types of regions. Whereas previous policy efforts have mostly concentrated on supporting knowledge-creation activities as the chief mechanisms to achieve higher economic performance, and implicitly on knowledge-intensive regions as prominent players in this regard, the results suggest that targeting all types of regions with dedicated innovation policies, as distinct from generic research policies, seems crucial to have them contribute fully to the achievement of the Europe 2020 strategy goals.

These results therefore directly speak to the current policy debate aimed at making Europe and its regions more competitive in international markets. In fact, from a policy perspective, the distinction among different innovation modes is of great importance in devising more targeted interventions able to act on specific local assets likely to lead to greater innovative and economic performances. This is the subject of the last section.

CONCLUSIONS

This paper has shown once more that studying productivity levels by looking at the effect of knowledge and innovation for the average region is not particularly informative because marked heterogeneity characterizes not only regional economic performance but also, and more relevantly, regional innovation behaviour. Although economic agents within each region may show a large variety of innovative behaviours (SRHOLEC and VERSPAGEN, 2012), the need to direct and orient innovation strategies at the regional level calls for the identification of the most important leverages of economic efficiency in regions showing some commonalities in their innovation modes. Therefore, accounting for the variety of regional innovation patterns is extremely important if the differentiated pathways from knowledge and innovation to productivity gains are to be fully grasped.

Interestingly, the key drivers of productivity gains vary across regions with their specific modes of innovation. Whereas scientific knowledge is relatively more prominent in regions with a stronger scientific knowledge endowment, knowledge embedded in capabilities is more desirable to achieve higher productivity levels in regions with a more limited scientific knowledge base. This result is consistent with, and further qualifies, previous findings (e.g., STERLACCHINI, 2008; CHARLOT *et al.*, 2012) indicating that inventive, and, indirectly, economic performance in less R&D-intensive regions is more closely linked to knowledge embodied in human capital (i.e., capabilities in the present case) than to additional investments in research and inventive activities. In some cases, therefore, a policy supporting R&D can prove to be extremely valuable, while it may produce virtually little, if not nil, effect in regions where the path to innovation is not based on the development of an internal scientific knowledge base.

Moreover, productivity levels arising from scientific knowledge are uneven and unbalanced across space, suggesting that sectoral policies like R&D should be carefully devised and adapted to local context conditions when translated into a regional setting. Interestingly, productivity levels stemming from innovation show a more diffusive and inclusive spatial pattern, although threshold effects seem to take place, consistently with previous research (CHARLOT *et al.*, 2012; CAPELLO and LENZI, 2013c).

These findings directly concern the current policy debate on how to make Europe and its regions become a knowledge-based economy and achieve smart growth. In particular, they enter the discussion on smart specialization strategies and demonstrate that European innovation policies must move away from a thematically/regionally neutral and generic strategy and instead look for selective and regionally appropriate strategies. If innovation policies are to support modernization in *all* European regions, they must comply with the specificities and potentials of individual regions, avoiding the opposite risks of a dispersion of public resources in undifferentiated ways, on the one hand, and of a concentration of all resources in a few regions where traditional R&D policy is apparently bound to yield the highest returns, leaving other regions unsupported, on the other.

These results indirectly witness that each territorial pattern of innovation can lead to efficiency gains; therefore, innovation policies should be oriented to reinforce the already established innovation pattern in each region, by boosting effectiveness of accumulated knowledge, fostering new applications and diversification, enlarging and deepening the local knowledge base (CAMAGNI and CAPELLO, 2013).

The two key concepts of 'embeddedness' and 'connectedness' – put forward in the recent smart specialization debate (MCCANN and ORTEGA-ARGILÉS, 2014) – are a useful starting point. However, innovation policies shall adapt the two concepts to the specificities of each pattern of innovation, and look for ad-hoc interventions, appropriate for each single territorial innovation pattern, with the aim to reinforce the virtuous aspects that characterize each pattern, and increase each pattern's efficiency.

This general policy strategy should not be open to doubts or criticisms concerning the possible risk of locking-in regions into their traditional specialization, jeopardizing their specific resilience in a fast-changing economic environment. In fact, the smart innovation strategy assumes, in its application to each regional innovation pattern, an evolutionary attitude, targeting, suggesting and supporting local learning processes towards the detection of new needs, new creative applications and diversification of established technologies, new forms of blending knowledge advancements and local specialization, the discovery, and possibly the orientation, of future technological trends. Even 'jumps' over a different innovation pattern might be foreseen in some regional cases, even if, given the responsibility in the management of public money, policy-makers should better stick to strengthening the upgrading and diversification processes inside each single innovation pattern – the least risky process, and the most likely successful one.

NOTES

1. Objective 1 regions are defined as regions with a per capita gross domestic product (GDP) lower than 75% of the European Union average.
2. For further details on the variables used in the cluster analysis and the variables representing the key territorial distinctive traits of the different groups of regions, see CAPELLO and LENZI (2013b).
3. The constitutive terms of the dummy variables accounting for the different territorial patterns of innovation are not included in equations (3) to (5), for both conceptual and technical reasons. In fact, they should be interpreted as the effect of each pattern when knowledge and/or innovation are set at zero. By construction, however, territorial patterns of innovation (and their respective dummies) are conceptually defined for positive values of knowledge and innovation. Therefore, it was decided not to include them in the regression framework.
4. Results are robust to the use of different estimation method to obtain TFP (OLS, two-stage least squares – with one or two-period lag of independent variables as instruments, SAR).
5. The test on the difference between the three coefficients is not statistically significant and indicates that the null hypothesis of coefficients equality cannot be rejected.

REFERENCES

ANSELIN L. (1988) *Spatial Econometrics: Methods and Models.* Kluwer, Dordrecht.

ASHEIM B. T. and ISAKSEN A. (2002) Regional innovation systems: the integration of local 'sticky' and global 'ubiquitous' knowledge, *Journal of Technology Transfer* **27**, 77–86.

BALCONI M., BRUSONI S. and ORSENIGO L. (2010) In defense of a linear model, *Research Policy* **39**, 1–13.

BEAUDRY C. and SCHIFFAUEROVA A. (2009) Who's right, Marshall or Jacobs? The localization vs urbanization debate, *Research Policy* **38**, 318–337.

CAMAGNI R. and CAPELLO R. (2013) Regional innovation strategies and the EU regional policy reform: towards smart innovation policies, *Growth and Change* **44**, 355–389.

CAPELLO R. (2013) Territorial patterns of innovation, in CAPELLO R. and LENZI C. (Eds) *Territorial Patterns of Innovation. An Inquiry on the Knowledge Economy in European Regions*, pp. 129–150. Routledge, Oxford.

CAPELLO R., CARAGLIU A. and NIJKAMP P. (2010) Territorial capital and regional growth: increasing returns in knowledge use, *Tijdschrift voor Economische en Sociale Geografie* **102**, 385–405.

CAPELLO R. and LENZI C. (Eds) (2013a) *Territorial Patterns of Innovation. An Inquiry on the Knowledge Economy in European Regions.* Routledge, Oxford.

CAPELLO R. and LENZI C. (2013b) Territorial patterns of innovation in Europe: a taxonomy of innovative regions, *Annals of Regional Science* **51**, 119–154.

CAPELLO R. and LENZI C. (2013c) Territorial patterns of innovation and economic growth in European regions, *Growth and Change* **44**, 195–227.

CARLAW K. I. and LIPSEY R. G. (2003) Productivity, technology and economic growth: what is the relationship?, *Journal of Economic Survey* **17**, 457–495.

CHARLOT S., CRESCENZI R. and MUSOLESI A. (2012) *An 'Extended' Knowledge Production Function Approach to the Genesis of Innovation in the European Regions.* Working Paper No. GAEL 2012-06 (available at http://www.grenoble.inra.fr/Docs/pub/A2012/gael2012-06.pdf).

COHEN W. and LEVINTHAL D. A. (1990) Absorptive capacity: a new perspective on learning and innovation, *Administrative Science Quarterly* **35**, 128–152.

COMMISSION OF THE EUROPEAN COMMUNITIES (2010a) *Europe 2020. A Strategy for Smart, Suitable and Inclusive Growth*. COM (2010)2020. European Commission, Brussels.

COMMISSION OF THE EUROPEAN COMMUNITIES (2010b) *Regional Policy Contributing to Smart Growth in Europe*. COM(2010) 553. European Commission, Brussels.

COMMISSION OF THE EUROPEAN COMMUNITIES (2012) *Guide to Research and Innovation Strategies for Smart Specialisations (RIS3)*. Smart Specialization Platform, May. European Commission, Brussels (available at: http://s3platform.jrc.ec.europa.eu/c/document_library/get_file?uuid=a39fd20b-9fbc-402b-be8c-b51d03450946&groupId=10157).

COOKE PH., ASHEIM B., BOSCHMA R., MARTIN R., SCHWARTZ D. and TÖDTLING F. (Eds) (2011) *Handbook of Regional Innovation and Growth*. Edward Elgar, Cheltenham.

DETTORI B., MARROCU E. and PACI R. (2011) Total factor productivity, intangible assets and spatial dependence in the European regions, *Regional Studies* **46**, 1401–1416. DOI: 10.1080/00343404.2010.529288

DIEWERT E. (1987) Index numbers, in EATWELL J., MILGATE M. and NEWMAN P. (Eds) *The New Palgrave, A Dictionary of Economics*. Macmillan, London (available at http://www.dictionaryofeconomics.com/article?id=pde1987_X001107).

EASTERLY W. and LEVINE R. (2001) It's not factor accumulation: stylized facts and growth models, *World Bank Economic Review* **15**, 177–219.

EDGERTON D. (2004) The linear model did not exist. Reflections on the history and historiography of science and research in industry in the twentieth century, in GRANDIN K., WORMS N. and WIDMALM S. (Eds) *The Science Industry Nexus*, pp. 31–57. Science History Publ., Sagamore Beach, MA.

EUROPEAN COUNCIL (2000) *Presidency Conclusions Lisbon European Council*. 23–24 March 2000 (No. 100/1/00). European Council, Brussels.

FAGERBERG J., MOWERY D. C. and NIGHTINGALE P. (2012) Introduction: The heterogeneity of innovation – evidence from the Community Innovation Surveys, *Industrial and Corporate Change* **21**, 1175–1180.

FARRELL M. J. (1957) The measurement of productive efficiency, *Journal of the Royal Statistical Society* **120**, 253–281.

FORAY D. (2009) Understanding smart specialisation, in PONTIKAKIS D., KYRIAKOU D. and VAN BAVEL R. (Eds) *The Question of R&D Specialisation*, pp. 19–28. JRC, European Commission, Directoral General for Research, Brussels.

FORAY D., DAVID P. and HALL B. (2009) Smart specialisation – the concept, *Knowledge Economists Policy Brief* **9**, 1–5.

FRENKEN K, VAN OORT F. G. and VERBURG T. (2007) Related variety, unrelated variety and regional economic growth, *Regional Studies* **41**, 685–697.

GRILICHES Z. (1987) Productivity: measurement problems, in EATWELL J., MILGATE M. and NEWMAN P. (Eds) *The New Palgrave, A Dictionary of Economics*. Macmillan, London (available at http://www.dictionaryofeconomics.com/article?id=pde1987_X001773).

GRILICHES Z. (1995) *The Discovery of the Residual*. NBER WO No. 5348. National Bureau of Economic Research (NBER), Cambridge, MA.

GUISO L., SAPIENZA P. and ZINGALES L. (2008) Social capital as good culture. Alfred Marshall Lecture, *Journal of European Economic Association* **6**, 295–320.

HENDERSON V., LEE T. and LEE Y. (2001) Scale externalities in Korea, *Journal of Urban Economics* **49**, 479–504.

HOLLANDERS H., TARANTOLA S. and LOSCHKY A. (2009a) *Regional Innovation Scoreboard (RIS) 2009*. Pro Inno Europe Paper No. 14. Enterprise and Industry Magazine, Brussels (available at: http://www.proinno-europe.eu/page/regional-innovation-scoreboard).

HOLLANDERS H., TARANTOLA S. and LOSCHKY A. (2009b) *Regional Innovation Scoreboard (RIS) 2009 – Methodology Report* (available at: http://www.proinno-europe.eu/page/regional-innovation-scoreboard).

HONG S., OXLEY L. and MCCANN P. (2012) A survey of innovation surveys, *Journal of Economic Surveys* **26**, 420–444.

IAMMARINO S. (2005) An evolutionary integrated view of regional systems of innovation: concepts, measures and historical perspectives, *European Planning Studies* **13**, 497–519.

JENSEN M. B., JOHNSON B., LORENZ E. and LUNDVALL B. A. (2007) Forms of knowledge and modes of innovation, *Research Policy* **36**, 680–693.

MCCANN P. and ORTEGA-ARGILÉS R. (2014) The role of the smart specialisation agenda in a reformed EU cohesion policy, *Scienze Regionali – Italian Journal of Regional Science* **13**, 15–32.

MADSEN J. (2008) Economic growth, TFP convergence, and the world export of ideas: a century of evidence, *Scandinavian Journal of Economics* **110**, 145–167.

MARROCU E. and PACI R. (2011) Education or creativity: what matters most for economic performance?, *Economic Geography* **88**, 369–401.

MARROCU E., PACI R. and USAI S. (2013) Productivity growth in the Old and New Europe: the role of agglomeration externalities, *Journal of Regional Science* **53**, 418–442.

MOWERY D. C. and ROSENBERG N. (1998) *Path of Innovation: Technological Change in 20th Century in America*. Cambridge University Press, Cambridge, MA.

NELSON R. R. (1959) The simple economics of basic scientific research, *Journal of Political Economy* **67**, 297–306.

ORGANISATION FOR ECONOMIC CO-OPERATION AND DEVELOPMENT (OECD) (2001) *OECD Productivity Manual*. OECD, Paris.

ORGANISATION FOR ECONOMIC CO-OPERATION AND DEVELOPMENT (OECD) (2010) *Typology of Regional Innovation Systems*. 20th Session of the Working Party on Territorial Indicators. OECD, Paris.

ORGANISATION FOR ECONOMIC CO-OPERATION AND DEVELOPMENT (OECD) (2011) *Territorial Outlook*. OECD, Paris.

PACI R. and MARROCU E. (2013) Knowledge assets and regional performance, *Growth and Change* **44**, 228–257.

PACI R. and USAI S. (2008) Agglomeration economies, spatial dependence and local industry growth, *Revue d'Economie Industrielle* **123**, 87–109.

PIKE A. and TOMANEY J. (1999) The limits to localization in declining industrial regions? Trans-national corporations and economic development in Sedgefield borough, *European Planning Studies* **7**, 407–428.

QUATRARO F. (2010) Knowledge coherence, variety and economic growth. Manufacturing evidence from Italian regions, *Research Policy* **39**, 1289–1302.

SOLOW R. M. (1956) A contribution to the theory of economic growth, *Quarterly Journal of Economics* **70**, 65–94.

SOLOW R. M. (1957) Technical change and the aggregate production function, *Review of Economics and Statistics* **39**, 312–320.

SRHOLEC M. and VERSPAGEN B. (2012) The voyage of the Beagle into innovation: explorations on heterogeneity, selection, and sectors, *Industrial and Corporate Change* **21**, 1221–1253. DOI: 10.1093/icc/dts026

STERLACCHINI A. (2008) R&D, higher education and regional growth: uneven linkages among European regions, *Research Policy* **37**, 1096–1107.

WINTJES R. and HOLLANDERS H. (2010) *The Regional Impact of Technological Change in 2020.* Report to the European Commission, Directorate General for Regional Policy, on behalf of the network for European Techno-Economic Policy Support (ETEPS AISBL) (available at: http://ec.europa.eu/regional_policy/sources/docgener/studies/pdf/2010_technological_change.pdf).

Do Technology Leaders Deter Inward R&D Investments? Evidence from Regional R&D Location Decisions in Europe

RENÉ BELDERBOS†‡§ and DIETER SOMERS¶

†*Department of Managerial Economics, Strategy and Innovation, Faculty of Economics and Business, University of Leuven, Leuven, Belgium*

‡*UNU-MERIT, Maastricht, the Netherlands*

§*School of Business and Economics, Maastricht University, Maastricht, the Netherlands*

¶*Department of Managerial Economics, Strategy and Innovation, Faculty of Economics and Business, University of Leuven, Antwerp, Belgium*

Belderbos R. and Somers D. Do technology leaders deter inward R&D investments? Evidence from regional R&D location decisions in Europe, *Regional Studies*. This paper examines the influence of local technology leaders on the attractiveness of regions for inward research and development (R&D) investments in an analysis of location choices for 196 cross-border R&D investments in EU-15 countries, 2003–08. While the strength of the regional technology cluster attracts inward R&D investments, investors are discouraged by a concentration of technology activities due to the presence of regional technology leaders. This effect occurs primarily if leaders organize their R&D in the region to reduce knowledge outflows to collocated firms – by drawing on intra-firm cross-border knowledge flows and complementary contributions of multiple R&D units.

Belderbos R. and Somers D. 技术领先者是否吓阻了向内的研发投资？欧洲区域研发的区位选择之证据，*区域研究*。本文在分析欧盟十五个成员国于2003年至2008年间的一百九十六件跨领域研发（R&D）投资的区位选择中，检视地方技术领先者，对于区域吸引向内研发投资的影响。区域技术集群虽然吸引了内向的研发投资，投资者却因区域技术领先者的存在所产生的技术活动集中而受到吓阻。此一影响，主要发生于领先者透过运用企业内的跨境知识流动及多重研发单位的互补性贡献，组织区域中用以降低知识外流的研发活动来配置企业之时。

Belderbos R. et Somers D. Les leaders en matière de technologie, est-ce qu'ils dissuadent les investissements en R et D? Des résultats des décisions d'emplacement régionales quant à la R et D en Europe, *Regional Studies*. À partir d'une analyse des choix d'emplacement auprès de 196 investissements en recherche et développement transfrontaliers (R et D), cet article examine l'influence des leaders locaux en matière de technologie quant à l'attractivité des régions pour l'investissement étranger en R et D dans les pays EU-15 (à savoir, les 15 pays membres de l'Union européenne à compter du 31 décembre 2003), entre 2003 et 2008. Alors que la puissance du cluster technologique régional attire les investissements étrangers en R et D, les investisseurs sont dissuadés par une concentration des activités technologiques à cause de la présence des leaders régionaux en matière de technologie. Cet effet se produit principalement si les leaders organisent leur R et D dans la région de façon à réduire les sorties de connaissances à destination des entreprises colocalisées – en bénéficiant des flux de connaissances transfrontaliers intra-entreprises et des contributions complémentaires des unités de R et D multiples.

Belderbos R. und Somers D. Wirken Technologieführer abschreckend auf ausländische FuE-Investitionen? Evidenz von den Entscheidungen über regionale FuE-Standorte in Europa, *Regional Studies*. In diesem Beitrag untersuchen wir mithilfe einer Analyse der Standortwahl für 196 grenzübergreifende Investitionen für Forschung und Entwicklung (FuE) in den EU-15-Ländern im Zeitraum von 2003 bis 2008 den Einfluss von lokalen Technologieführern auf die Attraktivität von Regionen für FuE-Investitionen aus dem Ausland. Während die Stärke des regionalen Technologie-Clusters ausländische FuE-Investitionen anzieht, wirkt eine Konzentration von technologischen Aktivitäten aufgrund der Präsenz regionaler Technologieführer auf

Investoren abschreckend. Dieser Effekt tritt vor allem auf, wenn die Technologieführer ihre FuE-Tätigkeit in der Region durch die Nutzung von firmeninternen grenzüberschreitenden Wissensströmen und ergänzenden Beiträgen verschiedener FuE-Einheiten so organisieren, dass die Wissensströme nach außen an andere Firmen am selben Standort verringert werden.

BELDERBOS R. y SOMERS D. ¿Disuaden los líderes tecnológicos a las inversiones extranjeras en I+D? Ejemplos de las decisiones de ubicación regional de la I+D en Europa, *Regional Studies*. A partir de un análisis de decisiones de ubicación para 196 inversiones transfronterizas en I+D en 15 países de la UE entre 2003 y 2008, en este artículo examinamos cómo influyen los líderes tecnológicos locales en el atractivo de las regiones para inversiones extranjeras en investigación y desarrollo (I+D). Si bien la fortaleza de la aglomeración tecnológica regional atrae inversiones extranjeras en I+D, los inversores están desmotivados por una concentración de actividades tecnológicas debida a la presencia de líderes tecnológicos regionales. Este efecto ocurre principalmente si los líderes organizan su I+D en la región para reducir las fugas de conocimiento a empresas asociadas, al basarse en flujos de conocimiento transfronterizos dentro de la empresa y contribuciones complementarias de varias unidades de I+D.

INTRODUCTION

Foreign multinational firms are responsible for a sizeable share of research and development (R&D) investments in most European countries (ORGANISATION FOR ECONOMIC CO-OPERATION AND DEVELOPMENT (OECD), 2007). Host country governments seek to attract these R&D investments since they provide numerous benefits to their region, such as highly skilled employment and knowledge spillovers to local firms (e.g. CARLSSON, 2006). Such knowledge spillovers resulting from clustered R&D activities favourably affect knowledge accumulation and economic growth of regions (AUDRETSCH and FELDMAN, 1996) and attract R&D investments by firms seeking to benefit from local knowledge networks (BELDERBOS *et al.*, 2009; CANTWELL and PISCITELLO, 2005; DEL BARRIO-CASTRO and GARCÍA-QUEVEDO, 2005).

Recent studies have pointed out that the presence in R&D clusters is not beneficial for all firms (ALCÁCER and ZHAO, 2012). While in general the agglomeration of R&D generates potential benefits due to increased knowledge spillovers, for technological leading firms there is a potentially important downside. Although incoming knowledge flows provide benefits, firms also face the risk that firm-specific valuable knowledge and technologies spill over to collocated firms, potentially undermining efforts to appropriate technologies and to earn returns on internal R&D efforts. The strong technology position of technologically leading firms creates an asymmetry: they have less to learn from incoming knowledge spillovers and at the same time face greater risks that valuable knowledge and frontier technologies are appropriated by neighbouring firms.

Although prior studies have confirmed the role of heterogeneity in the benefits of collocation in empirical models, by showing that leaders are less likely to choose agglomerated areas for new investments (ALCÁCER and CHUNG, 2007; CHUNG and ALCÁCER, 2002; SHAVER and FLYER, 2000), the position and strategies of incumbent leading firms with existing R&D facilities have not received due attention. The asymmetry in knowledge spillovers implies that technology leaders with existing R&D facilities have to face the possibility that firms set up new R&D establishments in their vicinity to benefit from knowledge spillovers. To prevent this, leading firms are likely to have incentives to discourage new R&D establishments in their locality. The extant literature has suggested that firms indeed employ a number of strategies in attempts to limit knowledge outflows, such as limiting personnel mobility by using non-compete clauses (e.g. MARX, 2011; MARX *et al.*, 2009), aggressive patent enforcement strategies (e.g. AGARWAL *et al.*, 2009), and geographically dispersed but interrelated internal R&D activities that ensure that technological knowledge developed in one location is of little use without complementary knowledge developed in other locations (ALCÁCER and ZHAO, 2012; ZHAO, 2006). Such strategies, while often intended to prevent knowledge flows to collocated firms, will also render new R&D investments less attractive to firms wishing to benefit from knowledge spillovers.

This paper addresses this issue by examining whether concentration of technological activities by local technology leaders displaying knowledge protection strategies discourages inward R&D investments, controlling for other relevant locational features. The focus here is on knowledge internalization strategies as a potential means to prevent outgoing spillovers. By utilizing information on local and cross-border self-citations in technology leaders' patents, this paper examines whether the dependence of local inventions on the leaders' internal technology development activities, and in particular on internal R&D abroad, makes regions relatively less attractive for investments. Since the presence of industry and regional R&D leaders has generally be seen as encouraging entry through agglomeration effects (AGRAWAL and COCKBURN, 2003), a potentially discouraging effects of the presence of technology leaders on multinational firms' R&D location choices provides an important qualification to such assessments – with related policy implications.

This article examines the role of technology leaders in an R&D location study at the regional level across multiple countries in the EU-15. Most previous R&D studies have been conducted at the national level (e.g. BELDERBOS *et al.*, 2009; KUMAR, 2001) or focused on regions within one country (AUTANT-BERNARD, 2006; HILBER and VOICU, 2010). It has however been shown that multinational firms take regions across multiple countries into consideration when they decide to locate foreign R&D investments (McCANN and MUDAMBI, 2004; THURSBY and THURSBY, 2009). This paper analyses the location of 196 foreign R&D investment projects drawing on data provided by the Financial Times' *fDi Markets Database* (2003–08). It examines firms' location choices at the regional (NUTS-1) level and identifies the strength of industry technology leaders as dominant players in regional R&D from the concentration of regional patenting activity among local firms. Investment locations and the relationship with technology clusters and technology leaders are examined at the relevant industry level. Empirical results suggest that while the strength and quality of local relevant technology development activities attract cross-border R&D investments, investors are discouraged by concentrated technology development activities due to the dominance of regional industry leaders. This discouraging effect occurs specifically in case the leaders exhibit an internally oriented R&D organization drawing on cross-border intra-firm knowledge flows in their technology development activities.

These results challenge the common premise in the literature on clusters (e.g. COOKE, 2001; PORTER, 2000; STORPER, 1995) that firms will generally benefit from knowledge spillovers when locating in regions with clustered industrial activity. The study highlights that geographical clustering and knowledge spillovers do not necessarily co-occur, but that knowledge spillovers are conditional on incumbent firms' strategic behaviour.

LITERATURE REVIEW AND PROPOSITIONS

This section provides a brief review of two relevant literature streams: the literature on agglomeration economies and firms' location decisions, and the literature on strategies to prevent knowledge spillovers. It proceeds by suggesting a base proposition on the negative relationship between concentrated technology ownership and location choice, after which two propositions on the effects of knowledge internalization strategies of technology leaders are derived.

Agglomeration economies

Agglomeration theory has been a dominant theory in explaining why firms tend to co-locate in the same region. The theory argues that firms can increase their profitability by locating in the proximity of other economic activities and related production facilities due to positive externalities that might stem from industry localization (MARSHALL, 1920). Several studies have shown that agglomeration economies are indeed an important factor in explaining foreign multinational location choices (e.g. BELDERBOS and CARREE, 2002; CHANG and PARK, 2005). MARSHALL (1920) highlighted three positive externalities that favour geographic concentration of industries: (1) industry demand that creates a pool of specialized labour, (2) industry demand that creates a pool of specialized inputs, and (3) knowledge spillovers among firms in an industry. Of all activities in the value chain, R&D activities benefit the most from knowledge spillovers among competing firms and consequently show the highest level of concentration (ALCÁCER, 2006; AUDRETSCH and FELDMAN, 1996; McCANN and MUDAMBI, 2004).

Empirical studies have shown that firms can improve their innovative performance by benefitting from knowledge spillovers in clusters (e.g. BEAUDRY and BRESCHI, 2003), although it is worth noting that some studies obtained contrasting or no effects in this regard (e.g. LEE and LIM, 2001). Less attention has been given to the notion that firms not only benefit from knowledge spillovers, but also that they contribute to these externalities. Knowledge outflows may hinder the firm from appropriating the value from its own inventions. Furthermore, as firms are heterogeneous, they will differ in the net benefits they receive from agglomeration economies (ALCÁCER and CHUNG, 2014; CHUNG and KALNINS, 2001; SHAVER and FLYER, 2000). Firms with the best technological capabilities, employees, distributors or suppliers will gain little from incoming spillovers, but may competitively suffer when their employees are hired by other firms and information on their technologies spills over to competing firms. In this context, a number of studies have found that larger or more R&D-intensive firms are less likely to invest in agglomerated areas (e.g. SHAVER and FLYER, 2000). However, once technology leaders have established their R&D facilities, they will have to face the possibility that other firms set up new R&D establishments in their vicinity to benefit from knowledge spillovers and they cannot easily avoid collocation. This is likely to provide leading firms with the incentives to adopt knowledge protection strategies discouraging new R&D establishments in their vicinity.

Strategies to prevent knowledge spillovers

Technology leaders can make use of several strategies to prevent outgoing knowledge spillovers from their existing R&D facilities. In general, one can distinguish between two types of knowledge protection strategies (De Faria and Sofka, 2010). One type relies on organizational processes within the firm to prevent knowledge spillovers, while the other relies on legal forms of protection based on formal applications or contracts.

Firms can adopt specific organizational practices to limit the risk of outgoing knowledge spillovers. As most technology leaders are large companies with R&D activities in multiple locations, they can take advantage of their geographically dispersed, but closely integrated, innovation networks. Studies have shown that technology leading firms can make use of strong internal linkages across multiple locations to leverage knowledge for competitive advantage, without risking critical knowledge outflows to competing firms (Alcácer and Zhao, 2012; Feinberg and Gupta, 2004; Zhao and Alcácer, 2007). Feinberg and Gupta (2004), for instance, documented that firms can minimize outgoing knowledge spillovers by integrating their global network of subsidiaries. As specialized complementary assets are essential for the commercialization of an invention (Teece, 1986), it follows that competing firms that lack such integration mechanisms are unable to benefit from the incoming knowledge spillovers (Zhao, 2006). Technology leading firms may also reduce outgoing knowledge flows by allocating less vulnerable projects to industrial clusters, by incorporating local innovations quickly into its global knowledge base and by the use of cross-cluster teams to strengthen control over locally developed technologies. The finding by Alcácer and Zhao (2012), that firms indeed make use of these mechanisms when surrounded by direct competitors, shows that the internal organization of R&D is often part of a well-considered strategy to limit outgoing knowledge transfers. Given that these cross-cluster R&D teams can be costly to manage (Frost *et al.*, 2002), firms will implement an internal linkage strategy when the benefits outweigh the costs, which is more likely under the threat of knowledge dissipation.

Another means to prevent harmful knowledge spillovers is the use of legal instruments such as patents and contracts. On the one hand, outgoing knowledge spillovers may be difficult to prevent through legal mechanisms, because knowledge spillovers are mostly tacit and often occur through interpersonal interactions. On the other hand, studies have documented that firms effectively make use of legal instruments aiming to reduce employee mobility (Almeida and Kogut, 1999; Rosenkopf and Almeida, 2003; Song *et al.*, 2003) and that non-compete contracts and employers' patent litigiousness significantly reduce employee mobility (Agarwal *et al.*, 2009; Marx *et al.*, 2009).

Propositions

The extant literature suggests that technology leading firms can effectively use mechanisms to prevent outgoing knowledge spillovers. While prior studies have generally seen the presence of technology leading firms as encouraging entry (e.g. Agrawal and Cockburn, 2003), this paper examines whether the presence of regional technology leading firms could have a potential entry deterring effect on foreign firms' R&D location choices. As firms take the potential for incoming knowledge spillovers into account when they locate their R&D facilities (Cantwell and Piscitello, 2005; Chung and Alcácer, 2002), they may refrain from locating their R&D activities in regions where technology leading firms have incentives to prevent outgoing knowledge flows.

There are various reasons why the actual opportunities to benefit from spillovers may be reduced if there is a strong concentration of technology activities in one or a few firms (McCann and Mudambi, 2005). Dispersion of R&D activities in a region among a larger number of R&D performing firms may fail to provide greater spillovers for prospective investors even in the absence of specific strategic behaviour to prevent outgoing knowledge flows. Chinitz (1961), for instance, contended that networking intensity between firms will be lower in highly concentrated industry structures. A smaller dispersion of R&D activities will also be associated with less diversity and a poorer content of the regional knowledge base (Feinberg and Gupta, 2004), rendering a close match between the knowledge generated by incumbents and the knowledge base of entrants less likely (Alcácer and Chung, 2014). Cantwell and Mudambi (2011) similarly suggest that industrial concentration reflects a relative closeness of the local innovation system. Finally, studies in the entrepreneurship literature suggest that the dispersion of technology activities among smaller incumbent firms might provide greater agglomeration benefits (Chen and Hambrick, 1995; Rosenthal and Strange, 2003), as small firms tend to be more open and entrepreneurial (Baum and Oliver, 1991).

It can be concluded that there is a range of arguments in prior studies suggesting that concentrated technology activities in regions provide fewer knowledge spillover benefits to entrants, such that these regions are less likely to attract R&D investments. This leads to the following baseline proposition:

Baseline proposition 1: A high concentration of technology activities in a region among incumbent local firms reduces the probability that foreign R&D investors locate in the region.

The negative correlation between concentration of technology ownership and the attractiveness of regions for R&D projects may vary, or depend on, the presence of technology protection strategies of incumbents. This paper focuses on R&D internalization strategies used by technology-leading firms to prevent valuable knowledge outflows (ALCÁCER and ZHAO, 2012; ZHAO, 2006).[1] These strategies imply that firms build on their own internal knowledge stock and complementary knowledge assets when developing new technologies. By developing technologies that require complementary knowledge and resources which are not readily available to potential imitators, technology leading firms can limit valuable knowledge outflows, preventing imitation (ALCÁCER and ZHAO, 2012; TEECE, 1986). For instance, JAMES (2014) finds a negative relationship between the patent self-citation ratio of technology leaders and the number of patent applications by competing firms. ALCÁCER and ZHAO (2012) show that technology leading firms are more often making use of internal linkages when they are surrounded by rival firms and suggest that this behavior is part of a strategy to prevent outgoing knowledge flows and to enhance appropriation. Indeed, adoption of such strategies was correlated with reduced citations of other firms to the focal firm's patent base. ZHAO's (2006) finding that firms are internalizing their technologies developed in countries with weak intellectual property rights (IPR) protection to a greater extent provides additional evidence that this internalization is part of a well-considered strategy to prevent knowledge expropriation.[2] If firms succeed in R&D internalization strategies, this can enhance competitive advantage (HALL *et al.*, 2003) and increase the firm's lead-time over competitors (COHEN *et al.*, 2000).

In summary, R&D internalization strategies as indicated by the importance of intra-firm knowledge flows are expected to reduce knowledge spillovers. As investing firms take the potential for incoming spillovers into account when they decide on the location of their foreign R&D activities, it follows that firms are discouraged to enter regions in which the leading firm adopted such strategies:

> *Proposition 2: A strong degree of internalization of technology development activities by local technology leaders reduces the probability that foreign R&D investors locate in the region.*

Regional technology leaders are often multinational firms operating multiple dispersed R&D facilities across a variety of countries. This allows them to assimilate, generate and integrate knowledge on a worldwide basis (FROST *et al.*, 2002; GHOSHAL and BARTLETT, 1990). It also offers the opportunity to internalize knowledge flows across borders (e.g. ZHAO, 2006). When technology leaders develop inventions that build upon prior internally developed technologies in R&D facilities abroad, this may present additional difficulties for competing firms to benefit from these leading firms' local R&D activities. Competing firms that seek to benefit from knowledge outflows and imitation strategies will face much greater difficulties to obtain needed complementary knowledge residing in several locations across national borders. This presents a much greater challenge than in case leading firms' internalization strategies are restricted to internal development of technologies in the same location. This leads to the next proposition:

> *Proposition 3: A strong degree of cross-border internalization of technology development activities by local technology leaders reduces the probability that foreign R&D investors locate in the region – more so than internalization strategies restricted to domestic technology development activities.*

DATA, VARIABLES AND METHODS

Information on cross-border R&D investments is obtained from the Financial Times' *fDi Markets Database* 2003–08. This database tracks global cross-border investments drawing on public information sources and information from local and national investment agencies. It includes information on the investing firms, the city and country of investment, the sector of investment, and the type of activity (R&D, manufacturing, logistics, distribution, etc.). Only information on R&D projects located in EU-15 countries is retrieved. Due to the limited availability of regional information required to construct the independent variables, in particular regional R&D intensity, the sample for analysis consists of 301 R&D investments made by 218 firms.[3] From among these projects, only those are selected for which the detailed press releases and company reports suggest that the mandate of the project extends to research. For these projects one can be sure that knowledge spillovers and knowledge sourcing are important, while projects focusing solely on development often imply limited activities focusing on application of technologies in the local context or design activities. Consequently, the analysis focuses on a sample of 196 R&D investments by 152 firms.[4] Firms based in the United States are responsible for the largest share of R&D projects (49%), followed by firms based in the UK (8%), France (8%), Germany (7%) and Japan (6%). Based on the information on the industry of the investment project, projects are categorized in 12 two-digit NACE (Nomenclature statistique des activités économiques dans la Communauté européenne) manufacturing industries, for which measures of industry-specific patent activity can be constructed (see below). The data set is complemented with information about the characteristics of the NUTS-1 regions in EU-15 countries using various data sources.

Dependent variable

The dependent variable is a binary variable, which indicates in which NUTS-1 region in the EU-15 the R&D investment is made. This variable takes the value 1 if a foreign firm made its R&D investment in host region

Table 1. Distribution of cross-border research and development (R&D) investments over NUTS-1 regions, 2003–08

NUTS-1 region	Frequency	Percentage	NUTS-1 region	Frequency	Percentage
Catalunya (ES5)	23	11.73	Baden-Württemberg (DE1)	2	1.02
Ireland (IE0)	21	10.71	Bayern (DE2)	2	1.02
East of England (UKH)	14	7.14	Berlin (DE3)	2	1.02
Flanders (FL)	11	5.61	Nordrhein-Westfalen (DEA)	2	1.02
South East England (UKJ)	9	4.59	Sachsen (DED)	2	1.02
Scotland (UKM)	9	4.59	North East Spain (ES2)	2	1.02
Denmark (DK0)	8	4.08	South Spain (ES6)	2	1.02
Comunidad de Madrid (ES3)	8	4.08	Finland (FI1)	2	1.02
Northern Ireland (UKN)	7	3.57	Central France (FR7)	2	1.02
Méditeranée (FR8)	6	3.06	North East England (UKC)	2	1.02
Bassin Parisien (FR2)	5	2.55	London (UKI)	2	1.02
North West Italy (ITC)	5	2.55	Wales (UKL)	2	1.02
Portugal (PT1)	5	2.55	Hamburg (DE6)	1	0.51
East Sweden (SE1)	5	2.55	South West France (FR6)	1	0.51
Île de France (FR1)	4	2.04	Isole (ITG)	1	0.51
Central Italy (ITE)	4	2.04	Luxembourg (LU0)	1	0.51
Walloon Region (BE3)	3	1.53	West Netherlands (NL3)	1	0.51
East France (FR4)	3	1.53	South Netherlands (NL4)	1	0.51
West France (FR5)	3	1.53	South Sweden (SE2)	1	0.51
South West England (UKK)	3	1.53	North Sweden (SE3)	1	0.51
East Austria (AT1)	2	1.02	North West England (UKD)	1	0.51
South Austria (AT2)	2	1.02	West Midlands (UKG)	1	0.51
West Austria (AT3)	2	1.02	Total	196	100

j; and 0 otherwise. The analysis includes 64 NUTS-1 regions.[5] Most R&D investments are made in the electronics and pharmaceutical industries with, respectively, 40% and 38% of the R&D investments.[6] The chemicals, transport and machinery industries are each responsible for about 5% of the investments, while the remaining industries are less well represented. No cross-border R&D investments are reported for the textiles, paper and mineral industries. Table 1 shows the distribution of the projects over NUTS-1 regions. It can be observed that most R&D investments are made in Catalunya (ES5) and Ireland (IE0) with 23 and 21 R&D projects respectively.

Technology concentration

To measure the level of concentration of technology development activities among firms based in the region, a Herfindahl–Hirschman index of company patent ownership is used.[7] This technology concentration index measures the concentration of patents among firms applied for by inventors based in the region and relevant to the industry. It is the sum of squares of the patent shares of individual firms with patents invented in the region and ranges between just over 0 (highly dispersed activities) and 1 (one company is responsible for all the region's relevant patents). Patent information from the European patent office is used to characterize technology concentration and technological strength (see below). The advantages of patent data are given by their consistent availability over time and their detailed information on technological content and location of inventive activity (GRILICHES, 1998). Patent data have been very frequently used in prior research on international R&D and as indicator of innovative activities (ALLRED and PARK, 2007; BELDERBOS, 2001; CANTWELL and PISCITELLO, 2005).

Patents are assigned to NUTS-1 regions based on the addresses of the inventors that are listed on the patents (DEYLE and GRUPP, 2005). In order to allocate patents to industries, the analysis utilizes a concordance table developed by SCHMOCH *et al.* (2003). This table links the Intellectual Patent Classification (IPC) technology codes of the patents to their corresponding NACE code at the two-digit level. If a patent lists multiple inventors and IPC classes, fractional counts are used to assign the patent to the region and industry. Patents are allocated to firms at the consolidated level including firm's majority-owned subsidiaries.

As these data operations, which had to be implemented across regions and industries, are likely to introduce measurement error in the data, a dummy variable was used to indicate whether the region is characterized by an above- or a below-the-median Herfindahl–Hirschman index rather than the detailed index itself. The dummy variable *high technology concentration* takes the value 1 if patent ownership in the industry and region is highly concentrated. Proposition 1 suggested that this variable has a negative effect on R&D location choice.

Internalization (patent self-citations)

To measure the internalization strategy of the regional technology leader, the patent self-citation ratio of this leader is calculated in the year prior to the investment. The technological leader is the firm that applied for the most patents in the NUTS-1 region and sector of the investing firm. The self-citation ratio is a measure of intra-firm knowledge transfers and represents the extent to which firms retain the value of R&D within internal boundaries (HALL *et al.*, 2001, 2003; JAMES and SHAVER, 2014; TRAJTENBERG *et al.*, 1997; ZHAO, 2006). The self-citation ratio is measured as the ratio of self-citations over total citations of a leading firm's patents invented in the region and related to the focal industry. A citation is considered a self-citation if the citing and cited patents are both assigned to the same firm (at the consolidated level).

To indicate combinations of technology concentration and internalization, three dummy variables are constructed. One dummy takes the value 1 if the region shows a strong (above median) technology concentration and if the technology leading firm is strongly internalizing its knowledge (the technology leader has an above median self-citation ratio). A second dummy variable takes the value 1 if the region exhibits strong technology concentration but when the technology leading firm is not strongly internalizing its knowledge flows (the firm has a below the median self-citation ratio). A third dummy variable takes the value 1 in case of weak technology concentration combined with strong internalization. The omitted reference dummy is the situation where there is neither strong technology concentration nor strong internalization. Proposition 2 suggests the strongest negative effect of the combination of high concentration and high internalization: strong local technology leaders with pronounced internalization strategies.

To test whether cross-border internalization strategies have a stronger effect than domestic internalization, cross-border and domestic self-citation ratios are calculated separately by identifying the country of the inventors of the cited patents. The foreign self-citation ratio is measured as the ratio of foreign self-citations over total citations of a leading firm's patents in the region and sector. The domestic self-citation ratio is calculated in a similar way by dividing domestic self-citations by total citations of all patents of the technology leader in the region and sector. Based on these ratios and the dummy variable identifying high technology concentration, seven combined dummy variables are constructed to examine what the roles are of technology concentration, domestic self-citations and foreign self-citations in discouraging R&D. The reference category is the case where a low technology concentration is combined with a low domestic and a low foreign self-citation ratio. Proposition 3 suggests the strongest negative effects in case of high technology concentration combined with a high foreign self-citation ratio, but less so for combinations of high concentration and high domestic self-citations.

Table 2 shows examples of regions with high technology concentration and high (foreign) self-citation ratios. Technology leaders are often large multinationals firms such as Bayer, Philips, Shell and Monsanto. These firms have large R&D budgets and operate R&D affiliates across many countries, opening up opportunities to internalize (cross-border) knowledge flows, which are reflected in high self-citation ratios. Still, there is ample heterogeneity in the intensity of self-citations, with, for example, AstraZeneca showing stronger internalization strategies than Bayer in the pharmaceutical industry.

Other explanatory variables

As control variables, a range of host region characteristics that have been found to influence R&D location choices in prior studies are included in the analyses. As not all location factors differ within a country across regions, or are available at the regional level, also included are a number of host country characteristics in addition to a full set of host country dummies.

The presence of clustered technology activities, proxied by patent activity, has been found to attract R&D investments in earlier studies (e.g. BELDERBOS *et al.*, 2009). The variable *technological strength* is included, which is the share of the region in the number of patent applications in the EU-15 allocated to the industry of the investing firms. It measures the availability of technological knowledge and potential R&D spillovers relevant for the industry. This variable has been constructed at the industry level to control for differences in the propensity to patent across industries. In addition, the analyses control for differences in *patent quality* across regions. The number of citations a patent receives (forward citations) mirrors the technological importance of the patent for the development of subsequent technologies and also reflects the economic value of inventions (HALL *et al.*, 2005; HARHOFF *et al.*, 2003; TRAJTENBERG, 1990). Information on forward citations received by the regional patents at the patent family level is collected and the variable *patent quality* is constructed: the average number of forward citations received by regional patents in an industry in the year prior to investment.[8] The *backward citation intensity of patents* invented in the region is also included to control for general characteristics of technology development in the region. This variable is measured as the average number of backward citations on patents in the NUTS-1 region and industry. Several studies have suggested that this backward citation frequency is an indicator of patent impact, as inventions using a wider array of prior art tend to be more valuable (LANJOUW and SCHANKERMAN, 2004). Finally, the empirical models include the *R&D intensity* of the host region (SHIMIZUTANI and TODO, 2008) to control for R&D spillovers at the broader regional level. This variable is the ratio of the total intramural R&D

Table 2. Regional technology leaders in selected regions with above-median technology concentration and leader's self-citation ratio

NUTS-1 region	Industry	HH index	Technology leader	Self-citation ratio	Foreign self-citation ratio
Walloon Region (BE3)	Food	1.00	Monsanto	1.00	1.00
Nordrhein-Westfalen (DEA)	Food	0.82	Bayer	1.00	0.33
Rheinland-Pfalz (DEB)	Chemicals	0.97	Basf	0.47	0.26
West Netherlands (NL3)	Chemicals	0.93	Shell	0.60	0.40
Thüringen (DEG)	Pharmaceuticals	1.00	Bayer	0.53	0.27
North West England (UKD)	Pharmaceuticals	0.99	Astrazeneca	0.73	0.73
Nord-Pas-de-Calais (FR3)	Rubber	1.00	Nexans	1.00	1.00
Walloon Region (BE3)	Rubber	1.00	Total	0.67	0.50
Catalunya (ES5)	Metals	0.50	Sandvik	1.00	1.00
Berlin (DE3)	Metals	0.63	Siemens	0.78	0.67
Portugal (PT1)	Machinery	0.81	Bosch	0.57	0.57
East Midlands (UKF)	Machinery	1.00	Siemens	0.41	0.41
Isole (ITG)	Electronics	1.00	STMicroelectronics	0.53	0.41
South Netherlands (NL4)	Electronics	0.95	Philips	0.52	0.39
South Sweden (SE2)	Transport	1.00	Ford	0.60	0.40
East of England (UKH)	Transport	0.78	Nissan	0.80	0.80

expenditure in the business sector of the host region over total GDP of that region.

Several studies have also found that academic research has a positive impact on the innovative performance of firms (e.g. COHEN *et al.*, 2002; FLEMING and SORENSON, 2004; MANSFIELD, 1995). As research has shown that these effects are highly localized (e.g. DEL BARRIO-CASTRO and GARCÍA-QUEVEDO, 2005; PIERGIOVANNI and SANTARELLI, 2001), firms may locate in the close vicinity of academic research institutions. The variable *academic research strength* is included, which is the share of the region in the number of university publications in the EU-15 allocated to the industry of the investing firms. University publications are counted at the level of regions and science fields and measure relevant publications for investing firms by linking science fields to the industry that firms are active in (BELDERBOS *et al.*, 2009).

The likelihood that a host region attracts foreign R&D investments may also rise due to market size and market sophistication (BARRELL and PAIN, 1996). The region's *population* and GDP per capita are included to account for respectively the market size and regional purchasing power and market sophistication. The regional *unemployment rate* is also included, since general labour availability can be an important factor in the decision to invest in a region (FRIEDMAN *et al.*, 1992). *Airport traffic* measures the total passengers embarked and disembarked on regional airports and is a measure of transportation infrastructure and regional connectivity, which may be particularly important for multinational firms (BEL and FAGEDA, 2008).

The models also include a number of investor-related characteristics. Dummy variables are included which take the value 1 if the firm had previously located an R&D investment in the region, or a non-R&D related investment, respectively. Manufacturing and sales investments often are a precursor to R&D investments to adapt processes and products to the local market (e.g. BELDERBOS, 2003), while prior R&D may lead to intra-firm collocation advantages. The *geographical distance* between the source city of the investing firm and the NUTS-1 region is included as a larger distance between the source and destination country can impede R&D investments due to increased informational uncertainty and coordination costs (CASTELLANI *et al.*, 2011; GHEMAWAT, 2001). *Language similarity* has also been shown to be an important factor for foreign R&D investments because it enables a smooth communication and coordination, lowering transaction costs (HEJAZI and SAFARIAN, 2002). A dummy variable is included that equals 1 if the source city and NUTS-1 region share the same official language.

In addition to host country dummies, the analysis includes two variables at the host country level. The *wage costs* of skilled R&D personnel (BELDERBOS *et al.*, 2009; KUMAR, 2001) is measured at the country level, as data at the regional level are not available. Data on average gross annual earning of qualified engineers drawing on UBS Price & Earning reports are used. The *b-index* is a measure for the tax pressure on R&D activities (WARDA, 2006). The higher this b-index, the less generous the tax treatment of R&D is in a particular country and the less attractive the country's regions might be for R&D investments.

To reduce variance all variables (except the binary variables) are transformed by taking their natural logarithm. All variables are lagged by one year with respect to the year of investment to allow a response time by the investing firms and to reflect the proper time-ordering.[9]

Methods

Within the location choice literature (e.g. ALCÁCER and CHUNG, 2007; HEAD *et al.*, 1995), the conditional logit model (MCFADDEN, 1974) has been widely used to analyse the location determinants of foreign direct investments. A drawback of this model is the restrictive assumption of independence of irrelevant alternatives (IIA). This property states that for any two alternatives the ratio of probabilities is independent of the characteristics of any other alternative in the choice set. Accordingly, the relative probability of any two alternatives is independent of the inclusion or removal of other alternatives. This characteristic also implies the absence of correlations between error terms across alternatives. This assumption is, however, frequently violated in location choice analyses. Recent studies (e.g. BASILE *et al.*, 2008; CHUNG and ALCÁCER, 2002) have therefore used the mixed logit model, which does not rely on the IIA assumption (MCFADDEN and TRAIN, 2000). As Hausman tests showed that this assumption was also violated in the sample, the analyses use mixed logit models.

These models start from a random utility maximization setting to examine location choices of R&D investments. Having a choice set of alternative host regions $r = 1,\ldots, R$ to locate an overseas R&D project at time t, a multinational firm f seeks to maximize its expected utility ($U_{\mathrm{fr},t}$) as a function of observable regional attributes and unobservable regional factors $\varepsilon_{\mathrm{fr}}$. The expected utility of a multinational firm f choosing region r among other host regions at time t can be expressed by the function:

$$U_{\mathrm{fr},t} = \alpha X_{\mathrm{fr},t-1} + \varepsilon_{\mathrm{fr}} \quad (1)$$

where $X_{\mathrm{fr},t-1}$ represents a vector of region-specific characteristics that can vary across industries or firms; and $\varepsilon_{\mathrm{fr}}$ defines a region-specific independent random disturbance term. While the standard conditional logit model restricts the coefficients α to be equal across firms, the mixed logit allows the coefficients to follow a distribution function. Accordingly, coefficients are decomposed into a fixed part and a random part that accounts for unobservable effects. The error term incorporates the random components of the coefficients and takes the following form:

$$\varepsilon_{\mathrm{fr}} = \lambda_f Z_{\mathrm{fr},t-1} + \mu_{\mathrm{fr}} \quad (2)$$

where $Z_{\mathrm{fr},t-1}$ is a vector of observable variables; while λ_{f} is a vector of randomly distributed parameters with zero mean following a normal distribution with variance Ω. The parameter μ_{fr} is an independent and identically distributed error term. If the parameter λ_{f} is observed, the probability that a firm f would locate its foreign R&D investment in city r could be expressed as a standard logit model. However, since the coefficients in the mixed logit model are not known but are assumed to follow a certain density function $g(\lambda_{\mathrm{f}})$, the locational choice probability has to be calculated over all possible values of λ_{f}. The mixed logit probability is therefore obtained by taking the integral of the multiplication of the conditional probability with the density functions describing the random nature of the coefficients. This is described by the following equation:

$$P_{\mathrm{fr}} = \int \frac{\exp(\alpha X_{\mathrm{fr},t-1} + \lambda_f Z_{\mathrm{fr},t-1})}{\sum_{j=1}^{J} \exp(\alpha X_{\mathrm{fj},t-1} + \lambda_f Z_{\mathrm{fj},t-1})} g(\lambda_{\mathrm{f}})\mathrm{d}(\lambda_{\mathrm{f}}) \quad (3)$$

The mixed-logit probability is essentially a weighted average of the logit formula evaluated at different values of the betas, where the weights are provided by the density function. In the current analysis, the most general approach is followed by allowing a continuous distribution function (BASILE *et al.*, 2008; BELDERBOS *et al.*, 2014b; CHUNG and ALCÁCER, 2002). As there is no closed form solution for the mixed logit probability, it has to be approximated by simulation techniques. The estimation of mixed logic models requires an assumption about the precise distribution function of the coefficients. The models in this paper are estimated under the assumption of a normal distribution – the distribution most commonly adopted as it has the most general properties. Other functional forms have less desirable properties or poor convergence levels (TRAIN, 2003). All regressions are run with 100 simulation draws (REVELT and TRAIN, 1998; TRAIN, 2003).

The random parameters in the regressions accommodate a rich array of differential preferences on the part of investors, while they also form the basis for accommodating correlations across alternative locations (a feature causing biased estimates in conditional logit models). Because there are no a priori expectations about whether certain variables should have a random component or not, the models allow the maximum number of variables to have both a fixed and a random component (BASILE *et al.*, 2008; CHUNG and ALCÁCER, 2002; REVELT and TRAIN, 1998).[10]

It is worth noting that the empirical model includes variables with different characteristic types. The sample includes a number of variables that vary over regions and time (e.g. GDP/cap, population), while there are also time-variant industry-specific variables at the region level (e.g. technological strength, academic research strength). Yet other factors are firm- and region-specific but remain constant over time (such as language similarity and geographic distance), while the variables firm's previous investments and previous R&D vary by investing firm, region and time. Finally, the sample includes also two variables that were only available at the country level: wage cost and the b-index.

EMPIRICAL RESULTS

Results of the mixed logit models are reported in Table 3. Model 1 includes the control variables only, while model 2 adds the technology concentration dummy (Proposition 1). In model 3, three dummy variables

Table 3. Mixed-logit analysis of European regional location choices for research and development (R&D) investment projects, 2003–08

	Model 1	Model 2	Model 3	Model 4
Technological Strength	0.838*	0.855*	1.177**	1.061**
	(0.429)	(0.457)	(0.480)	(0.525)
Academic Research Strength	0.695	0.366	1.138	0.639
	(1.397)	(1.329)	(1.449)	(1.631)
R&D Intensity	0.379	0.359	0.463	0.287
	(0.474)	(0.465)	(0.602)	(0.589)
Patent Quality	0.390	0.383	0.591**	0.615**
	(0.242)	(0.234)	(0.271)	(0.288)
Backward Citation Intensity	0.299	0.258	0.414	0.492
	(0.239)	(0.246)	(0.310)	(0.300)
GDP per Capita	−1.866**	−2.116***	−2.099**	−2.444***
	(0.763)	(0.774)	(0.850)	(0.851)
Population	0.171	0.255	0.431	0.468
	(0.412)	(0.411)	(0.458)	(0.443)
Firm's Previous Investments	0.005	0.852	0.161	−1.066
	(1.055)	(0.771)	(1.216)	(1.974)
Firms' Previous R&D	35.208***	31.809***	42.131***	69.863***
	(7.079)	(3.967)	(7.599)	(21.994)
Unemployment Rate	−1.164**	−1.022**	−1.037*	−1.355**
	(0.498)	(0.484)	(0.566)	(0.553)
Airport Traffic	0.717***	0.715***	0.699***	0.760***
	(0.192)	(0.194)	(0.204)	(0.218)
Geographic Distance	−0.158	−0.296	−0.039	−0.311
	(0.420)	(0.371)	(0.442)	(0.490)
Language Similarity	0.705**	0.636**	0.770**	0.635*
	(0.311)	(0.306)	(0.357)	(0.382)
Wage Cost	0.560	0.765	−0.239	0.726
	(1.313)	(1.482)	(1.466)	(1.660)
B-index	−2.776	−2.183	−2.997	−4.967*
	(2.420)	(2.396)	(2.758)	(2.858)
High Technology Concentration		−0.749*		
		(0.383)		
High Technology Concentration/High Self-citation			−4.889**	
			(2.163)	
High Technology Concentration/Low Self-citation			−0.009	
			(0.301)	
Low Technology Concentration/High Self-citation			−0.619	
			(0.712)	
High Technology Concentration/High Domestic Self-citation/High Foreign Self-citation				−26.012**
				(12.777)
High Technology Concentration/High Domestic Self-citation/Low Foreign Self-citation				−1.280
				(1.474)
High Technology Concentration/Low Domestic Self-citation/High Foreign Self-citation				−2.519***
				(0.749)
High Technology Concentration/Low Domestic Self-citation/Low Foreign Self-citation				−0.331
				(0.815)
Low Technology Concentration/High Domestic Self-citation/High Foreign Self-citation				−17.326
				(15.123)
Low Technology Concentration/High Domestic Self-citation/Low Foreign Self-citation				−0.379
				(0.931)
Low Technology Concentration/Low Domestic Self-citation/High Foreign Self-citation				0.245
				(0.350)
Random parts coefficients				
R&D Intensity				−1.945**
Population	1.501***	1.657***	2.178***	2.355***
Firm's Previous Investments	−5.193*	3.395**	6.279**	11.007*
Geographic Distance	1.520**	−1.903***	2.106***	−2.770***
Wage Cost		−2.384*		2.997**
B-index			−2.656**	
Average Forward Citations			−1.304**	−1.430*
High Technology Concentration/High Self-citation			−5.229***	
Low Technology Concentration/High Self-citation			3.346**	
High Technology Concentration/High Domestic Self-citation/High Foreign Self-citation				−20.081**

(Continued)

Table 3. Continued

	Model 1	Model 2	Model 3	Model 4
Number of observations	7559	7559	7559	7559
Number of projects	196	196	196	196
Number of investing firms	152	152	152	152
Number of regions (maximum)	64	64	64	64
Wald chi^2	154.2***	291.9***	178.2***	260.9***

Note: Error terms are cluster by investing firm. Significance levels: ***$p < 0.01$; **$p < 0.05$; *$p < 0.10$. Only significant random components of the coefficients are reported.

are added to examine whether the internalization strategy of the technology leading firm of the region significantly deters R&D investments (Proposition 2). Finally, model 4 introduces seven dummy variables to test whether cross-border internal linkages are more effectively discouraging R&D than domestic internal linkages (Proposition 3).

Most of the control variables have the expected sign. The technological strength of the NUTS-1 regions has a significantly positive influence on the propensity to locate R&D investments in a region. Patent quality also matters: its coefficient is always positive and turns significant in the more encompassing models 3 and 4. Other significant influences are language similarity, airport infrastructure, the unemployment rate and firm's previous R&D investments. The large coefficient of previous R&D investments shows that multinational firms have a clear tendency to invest in regions where they are already present. This points to the importance of internal agglomeration and collocation forces as documented in prior studies (ALCÁCER and DELGADO, 2013; DEFEVER, 2006). One unexpected finding is the significantly negative effect of regional GDP per capita. This negative effect may be caused by a potentially high correlation with an omitted variable in the model: regional wage costs. As wage costs have a negative impact on R&D location decisions, the correlated variable GDP per capita might absorb part of this effect. Due to the unavailability of data, it is not feasible to include regional wage variables in the empirical models and instead the paper includes wage costs at the national level. Another surprising result is the insignificance of academic research strength – although its sign is positive, as expected. Part of the explanation may be sought in the restricted sample of regions to those that exhibit patented inventions. As this restriction tends to eliminate regions with poorly developed academic infrastructure, the variation in the academic research variable is substantially reduced. In addition, the data requirements and focus of the analysis limit the time frame of analysis and number of investments that can be covered in the analysis, in comparison with other location studies (e.g. BELDERBOS *et al.*, 2014b).

Model 2 shows that the technology concentration dummy is significantly negative (at the 10% level), while the other coefficients are not strongly affected. A high technology concentration of technology development activities in a region reduces the probability that foreign R&D investors locate in that region, in support of Proposition 1. In model 3 only the coefficient on the dummy representing regions with both high technology concentration and high internalization is significantly negative, while the other two dummies are not significant. This suggests that it is not the concentration as such that distracts investments, but concentration accompanied with a strong internal orientation of R&D by the regional technology leading firm, in support of Proposition 2.

Finally, Model 4 introduces seven dummy variables to compare the effects of cross-border and domestic knowledge internalization. The reference category is low technology concentration in combination with low domestic and low cross-border internalization. The two significantly negative coefficients for regions with high technology concentration combined with high cross-border internalization (with or without high domestic internalization) clearly suggest that foreign R&D investors are deterred from investing in regions where technology leaders follow internalization strategies drawing on cross-border knowledge flows. These patterns suggest that it is cross-border internalization, rather than domestic internalization, that discourages investment, consistent with proposition 3.

Among the random terms of the coefficients, R&D intensity, population, firm's previous investments, geographic distance, wage cost and the b-index are significant, as well as three dummy variables testing the propositions. This indicates that there is unspecified heterogeneity in preferences for different sets of regions and among investors, which will lead to a violation of the IIA in standard conditional logit models. Likelihood ratio tests comparing these models with models having only fixed coefficients also show that these models are significantly different from each other, which underscores the need to utilize mixed logit models. Although it is difficult to assess what the underlying causes are of the observed heterogeneity in the effects of the independent variables, it can be noted that the findings are broadly consistent with previous location studies discussing heterogeneity in the valuation of location specific characteristics. For instance, BASILE *et al.* (2008) find variance in the importance of population, R&D intensity, wage costs, firm prior experience and distance; CHUNG and ALCÁCER (2002) emphasize the varying importance of R&D intensity.

SENSITIVITY ANALYSIS

The measure of knowledge internalization strategies can be correlated with broader strategic behaviour by technology leaders to reduce knowledge outflows. High self-citation ratios are usually correlated with a low external collaboration intensity, as firms engaged in collaboration networks or alliances share more knowledge compared with non-allying firms (GOMES-CASSERES *et al.*, 2006). To explore this, an imperfect proxy for external R&D collaboration by the leading firms was included. As indicator of collaboration, the share of co-patents in the leading firm's regional patents was used. It can be noted that this variable is rather imperfect, as only a fraction of R&D collaboration activities will eventually lead to co-patents, given the complex property rights implications of shared patent rights (BELDERBOS *et al.*, 2014a). This co-patent ratio was (weakly) negatively related to the self-citation ratio, as expected. Including the variable in the regressions resulted in a positive sign, but the coefficient was not significantly different from zero.

Another explorative analysis examined potential investor heterogeneity by excluding investing firms holding large patent portfolios and planning large R&D investments: in these cases it may be that investing firms are similarly afraid of generating outgoing spillovers to local firms (e.g. ALCÁCER and CHUNG, 2014), and local leaders' strategies might become less relevant. The resulting models however did not identify a clear difference for the sample cleaned of such 'leading investors'. Similarly, subsample analysis of investment projects limited to knowledge intensive industries did not indicate stronger effects of knowledge sourcing related variables, with the exception of large coefficients for technological strength.

CONCLUSIONS

If firms take the potential for incoming knowledge spillovers into account when they decide on the location of new R&D facilities (e.g. AHARONSON *et al.*, 2007; CHUNG and ALCÁCER, 2002), they may refrain from locating their R&D activities in regions where incumbent firms are making use of mechanisms to prevent outgoing knowledge flows. Leading technology firms in particular may have incentives to prevent knowledge spillovers, including those to new entrants, as they have most to lose from outgoing spillovers and have relatively less to gain from incoming knowledge. One way to mitigate risks of outgoing spillovers is to adjust the internal organization of R&D activities by focusing on knowledge internalization, with new technology development drawing on technological knowledge available within the firm's boundaries (ALCÁCER and ZHAO, 2012). In particular if technology development rests on complementary knowledge residing in several R&D locations across national borders, collocated firms are likely to have fewer opportunities to benefit from leading firms' local R&D efforts (ZHAO, 2006). This implies that concentration of technology development activities in a region in leading firms that follow R&D internalization strategies may reduce the probability that foreign R&D investors locate in the region.

This paper finds support for the notion that technology leaders discourage inward R&D investments in an analysis of the location of 196 foreign R&D investment projects drawing on data provided by the Financial Times' *fDi Markets Database* (2003–08). It analyses firms' location choices at the regional (NUTS-1) level and identifies the strength of industry technology leaders as dominant players in regional R&D from the concentration of regional patenting activity among local firms. The paper analyses investment locations and the relationship with technology clusters and technology leaders at the relevant industry level, and considers R&D investment projects across manufacturing industries. Empirical results show that, while the strength and quality of local relevant technology development attracts R&D projects by foreign firms, investors are discouraged by a concentration of technology development activities among regional industry leaders. It is shown that this effect depends on the strategy of leading local firms to internalize technology development activities (measured by the intensity of self-citations on firms' patents). By developing technologies that build upon internal knowledge stocks and that require complementary knowledge and resources which are not readily available to potential imitators, knowledge dissipation to, and imitation by, other firms remains limited (FEINBERG and GUPTA, 2004; TEECE, 1986). Consequently, these internalization mechanisms allow the technology leading firm to minimize outgoing knowledge spillovers that could endanger its competitive position (ALCÁCER and ZHAO, 2012). Furthermore, the analysis shows that internalization strategies focusing on intra-firm cross-border knowledge flows with R&D units abroad have a much stronger discouraging effect on inward R&D than domestic internal linkages. Cross-border integration of technology development is a greater impediment to knowledge spillovers to investing firms as effective knowledge sourcing may require these firms to combine their investment with complementary R&D activities of their own in other locations.

The findings in this paper contribute to the literature on agglomeration and location decisions (ALCÁCER and CHUNG, 2007; CHUNG and ALCÁCER, 2002; SHAVER and FLYER, 2000), by focusing on the role of incumbent strategies to avoid outgoing spillovers, which has not received due attention. This article contributes to a stream of research suggesting that concentration of ownership of local resources may render locations less attractive (e.g. CANTWELL and MUDAMBI, 2011; ALCÁCER and CHUNG, 2014) by highlighting that the presence of dominant firms in local R&D clusters discourages entry primarily if this dominance is accompanied by R&D organizational strategies to prevent knowledge spillovers. In this manner, the findings challenge the common premise in the literature on clusters (e.g. COOKE, 2001; PORTER, 2000; STORPER, 1995) that firms will generally benefit from knowledge spillovers when locating in regions with clustered industrial activity, and call for future research investigating more detailed properties of regional innovation systems as antecedents of or impediments to such knowledge spillovers.

The results suggest that the presence of 'regional champions' or 'anchor firms', while increasing innovation activities and output in a region (AGRAWAL and COCKBURN, 2003; FELDMAN, 2003), at the same time may discourage inward R&D investments in the region. If the leading firms organize their internal R&D to prevent knowledge spillovers, this can have detrimental effects for technology development in the region and accordingly has important consequences for regional innovation policies. Limited knowledge spillovers to other firms in the region will limit agglomeration benefits and positive externalities, potentially reducing knowledge accumulation and economic growth (AUDRETSCH and FELDMAN, 1996; BALDWIN and MARTIN, 2004; MARTIN and OTTAVIANO, 1999). At the same time, it reduces inward R&D investments, and policies to attract inward R&D competition may prove to be less effective. This suggests the importance of implementing policies to promote more open innovation strategies by local technology leaders. There are several ways to promote knowledge diffusion. Policy makers can take measures to improve local labour mobility and stimulate spin-offs and other open innovation initiatives, as these are major mechanisms to increase knowledge spillovers (ALMEIDA and KOGUT, 1999; CAPELLO, 1999; CAPELLO and FAGGIAN, 2005). A useful instrument may be targeted R&D subsidies that require collaboration with other firms, as R&D collaboration may alter the R&D internalization strategies and generate broader spillovers to collaboration partners (e.g. AUTANT-BERNARD *et al.*, 2007; GOMES-CASSERES *et al.*, 2006).

The analysis in this paper is subject to a number of limitations, which could provide avenues for further research. First, the analysis only take into account knowledge protection strategies through internal R&D organization. To the extent that incumbent firms may use a range of protection strategies simultaneously (e.g. patent litigation strategies and non-compete clauses), the findings on the importance of internalization strategies may partially pick up the effect of correlated protection strategies. Future work should aim to analyse the effects of multiple knowledge protection strategies. A stumbling block will be the lack of available data at the regional or firm level, for instance on labour mobility and non-compete clauses. Second, this study focuses on 'horizontal' knowledge spillovers due to agglomeration of firms in the same industry. Future studies should take the role of spillovers from upstream (suppliers) or downstream (customers) simultaneously into account. Most studies examining supplier and customer spillovers (FRITSCH and FRANKE, 2004; KLOMP and VAN LEEUWEN, 2001; TETHER, 2002) have relied on survey data, and the challenge will be to extend analysis in a representative manner to the level of regions and industries across countries.

Third, the analysis uses a relatively broad spatial disaggregation of European countries into regions, at the NUTS-1 level., which might not be necessarily capturing the most relevant R&D clusters. The lack of a regional delineation based on economic areas is a general problem of analysing regional development in Europe, where one has to work with existing regional administrative definitions (BOTTAZZI and PERI, 2003; MAURSETH and VERSPAGEN, 2002; TAPPEINER *et al.*, 2008). Although the literature on knowledge spillovers suggests that spillovers decay rapidly at distance (e.g. AUTANT-BERNARD, 2001; BELENZON and SCHANKERMAN, 2012; JAFFE *et al.*, 1993), the available estimates of decay functions do not directly suggest that NUTS-1 regions are particularly large in this respect. Nevertheless, a future challenge for research clearly is to work with 'economic areas' in Europe based on actual agglomeration and commuting patterns.

Acknowledgements – The authors would like to thank three anonymous reviewers, Srikanth Paruchuri, Georg Licht and Gary Dushnitsky, and the participants at the 2011 Strategic Management Society in Miami, Florida, the 2012 Innovation Research Workshop at Maastricht University, the 2012 Geography of Innovation Conference in Saint Etienne, the 2012 Academy of International Business conference in Washington, DC, and the 2013 European Strategy, Entrepreneurship & Innovation (SEI) Doctoral Consortium at Imperial College London for comments made on earlier drafts of the paper.

Disclosure statement – No potential conflict of interest was reported by the authors.

Supplemental data – Supplemental data for this article can be accessed at http://dx.doi.org/10.1080/00343404.2015.1018881

NOTES

1. The analysis is limited to strategies related to R&D organization within the leading firms, as strategies based on legal forms of protections (e.g. employment contracts and non-disclosure agreements) are difficult to identify.

2. Self-citations indicate the extent to which firms build on their own prior technology development efforts in developing new inventions. It is a measure of intra-firm knowledge transfer and represents the extent to which firms retain the value of R&D within internal boundaries (JAMES and SHAVER, 2014; TRAJTENBERG *et al.*, 1997; ZHAO, 2006).
3. Some variables (such as R&D Intensity) are not consistently reported by EUROSTAT for a number of countries across years, such that a number of region-years observations are omitted. Greece is missing due to lack of information on a range of regional characteristics, but Greece did not receive R&D investments during the period of observation. Since (apart from Greece) all host countries remain in the data set and the omission of specific years is not systematic, this forced attrition of the sample is unlikely to lead to estimation bias. The models also exclude regions and industry combinations for which no company patents are recorded, as for these regions the key explanatory variables related to technology concentration and internalization cannot be calculated. Regions without any company patents are typically smaller regions in less patent-intensive industries. It has been confirmed that this sample attrition has no appreciable impact on the results by exploring models setting the concentration and internalization variables to zero in case the region has no patented inventions.
4. In a robustness tests development projects are included in the analysis. This produces similar, though less pronounced, empirical results.
5. In case the investing firm is based in the EU-15, the regions in their home country are omitted from the choice set. As the FDI data only cover cross-border investments, R&D investments in the home country are not observed by definition.
6. Table A1 in the Supplementary data online shows the distribution of R&D projects over industries.
7. Patents owned by individuals, universities, hospitals and governmental non-profit organizations are excluded.
8. Due to the truncation of patents in the most recent years and for consistency, the window of forward citations is limited to two years.
9. Definitions and summary statistics of the variables are provided in Table A2 in the Supplementary data online. The correlation table is available from the authors upon request.
10. As the statistical program used (the mixlogit command in STATA) only allows 20 random variables, non-random coefficients are assumed for two variables. The authors have chosen two control variables with insignificant random parameters and for which preference heterogeneity ex-ante was less likely as fixed: *Previous R&D* and *Airport traffic*. Alternative settings for non-random coefficients produced consistent results.

REFERENCES

AGRAWAL A. and COCKBURN I. (2003) The anchor tenant hypothesis: exploring the role of large, local, R&D-intensive firms in regional innovation systems, *International Journal of Industrial Organization* **21**, 1227–1253. doi:10.1016/S0167-7187(03)00081-X

AGARWAL R., GANCO M. and ZIEDONIS R. H. (2009) Reputations for toughness in patent enforcement: Implications for knowledge spillovers via inventor mobility, *Strategic Management Journal* **30**, 1349–1374. doi:10.1002/smj.792

AHARONSON B. S., BAUM J. A. C. and FELDMAN M. P. (2007) Desperately seeking spillovers? Increasing returns, industrial organization and the location of new entrants in geographic and technological space, *Industrial and Corporate Change* **16**, 89–130. doi:10.1093/icc/dtl034

ALCÁCER J. (2006) Location choices across the value chain: how activity and capability influence collocation, *Management Science* **52**, 1457–1471. doi:10.1287/mnsc.1060.0658

ALCÁCER J. and CHUNG W. (2007) Location strategies and knowledge spillovers, *Management Science* **53**, 760–776. doi:10.1287/mnsc.1060.0637

ALCÁCER J. and CHUNG W. (2014) Location strategies for agglomeration economies, *Strategic Management Journal* **35**, 1749–1761. doi:10.1002/smj.2186

ALCÁCER J. and DELGADO M. (2013) *Spatial Organization of Firms and Location Choices through the Value Chain*. Working Paper. Harvard Business School, Boston, MA.

ALCÁCER J. and ZHAO M. (2012) Local R&D strategies and multilocation firms: the role of internal linkages, *Management Science* **58**, 734–753. doi:10.1287/mnsc.1110.1451

ALLRED B. B. and PARK W. G. (2007) Patent rights and innovative activity: evidence from national and firm-level data, *Journal of International Business Studies* **38**, 878–900.

ALMEIDA P. and KOGUT B. (1999) Localization of knowledge and the mobility of engineers in regional networks, *Management Science* **45**, 905–917. doi:10.1287/mnsc.45.7.905

AUDRETSCH D. B. and FELDMAN M. P. (1996) R&D spillovers and the geography of innovation and production, *American Economic Review* **86**, 630–640.

AUTANT-BERNARD C. (2001) Science and knowledge flows: evidence from the French case, *Research Policy* **30**, 1069–1078. doi:10.1016/S0048-7333(00)00131-1

AUTANT-BERNARD C. (2006) Where do firms choose to locate their R&D? A spatial conditional logit analysis on French data, *European Planning Studies* **14**, 1187–1208. doi:10.1080/09654310600933314

AUTANT-BERNARD C., MAIRESSE J. and MASSARD N. (2007) Spatial knowledge diffusion through collaborative networks, *Papers in Regional Science* **86**, 341–350. doi:10.1111/j.1435-5957.2007.00134.x

BALDWIN R. E. and MARTIN P. (2004) Agglomeration and regional growth, *Handbook of Regional and Urban Economics* **4**, 2671–2711. doi:10.1016/S1574-0080(04)80017-8

Barrell R. and Pain N. (1996) An econometric analysis of U.S. foreign direct investment, *Review of Economics and Statistics* **78**, 200–207. doi:10.2307/2109921

Basile R., Castellani D. and Zanfei A. (2008) Location choices of multinational firms in Europe: the role of EU cohesion policy, *Journal of International Economics* **74**, 328–340. doi:10.1016/j.jinteco.2007.08.006

Baum J. A. C. and Oliver C. (1991) Institutional linkages and organizational mortality, *Administrative Science Quarterly* **36**, 187–218.

Beaudry C. and Breschi S. (2003) Are firms in clusters really more innovative?, *Economics of Innovation and New Technology* **12**, 325–342. doi:10.1080/10438590290020197

Bel G. and Fageda X. (2008) Getting there fast: globalization, intercontinental flights and location of headquarters, *Journal of Economic Geography* **8**, 471–495. doi:10.1093/jeg/lbn017

Belderbos R. (2001) Overseas innovations by Japanese firms: an analysis of patent and subsidiary data, *Research Policy* **30**, 313–332. doi:10.1016/S0048-7333(99)00120-1

Belderbos R. (2003) Entry mode, organizational learning, and R&D in foreign affiliates: evidence from Japanese firms, *Strategic Management Journal* **24**, 235–259. doi:10.1002/smj.294

Belderbos R. and Carree M. (2002) The location of Japanese investments in China: agglomeration effects, *keiretsu*, and firm heterogeneity, *Journal of the Japanese and International Economies* **16**, 194–211. doi:10.1006/jjie.2001.0491

Belderbos R., Cassiman B., Faems D., Leten B. and Van Looy B. (2014a) Co-ownership of intellectual property: exploring the value-appropriation and value-creation implications of co-patenting with different partners, *Research Policy* **43**, 841–852. doi:10.1016/j.respol.2013.08.013

Belderbos R., Leten B. and Suzuki S. (2009) *Does Excellence in Academic Research Attract Foreign R&D?* UNU-MERIT Working Paper. Maastricht Economic and Social Research and Training Centre on Innovation and Technology, Maastricht.

Belderbos R., Van Roy V., Leten B. and Thijs B. (2014b) Academic research strengths and multinational firms' foreign R&D location decisions: evidence from R&D investments in European regions, *Environment and Planning A* **46**, 920–942. doi:10.1068/a45536.

Belenzon S. and Schankerman M. (2012) Spreading the word: geography, policy, and knowledge spillovers, *Review of Economics and Statistics* **85**, 884–903.

Bottazzi L. and Peri G. (2003) Innovation and spillovers in regions: evidence from European patent data, *European Economic Review* **47**, 687–710. doi:10.1016/S0014-2921(02)00307-0

Cantwell J. A. and Mudambi R. (2011) Physical attraction and the geography of knowledge sourcing in multinational enterprises, *Global Strategy Journal* **1**, 206–232. doi:10.1002/gsj.24

Cantwell J. and Piscitello L. (2005) Recent location of foreign-owned research and development activities by large multinational corporations in the European regions: the role of spillovers and externalities, *Regional Studies* **39**, 1–16. doi:10.1080/0034340052000320824

Capello R. (1999) Spatial transfer of knowledge in high technology milieux: learning versus collective learning processes, *Regional Studies* **33**, 353–365. doi:10.1080/00343409950081211

Capello R. and Faggian A. (2005) Collective learning and relational capital in local innovation processes, *Regional Studies* **39**, 75–87. doi:10.1080/0034340052000320851

Carlsson B. (2006) Internationalization of innovation systems: a survey of the literature, *Research Policy* **35**, 56–67. doi:10.1016/j.respol.2005.08.003

Castellani D., Palmero A. J. and Zanfei A. (2011) *The Gravity of R&D FDIs*. Working Paper Series in Economics, Mathematics and Statistics, WP-EMS No. 2011/06.

Chang S. J. and Park S. (2005) Types of firms generating network externalities and MNCs' co-location decisions, *Strategic Management Journal* **26**, 595–615. doi:10.1002/smj.464

Chen M. J. and Hambrick D. C. (1995) Speed, stealth, and selective attack: how small firms differ from large firms in competitive behavior, *Academy of Management Journal* **38**, 453–482.

Chinitz B. (1961) Contrasts in agglomeration: New York and Pittsburgh, *American Economic Review* **51**, 279–289.

Chung W. and Alcácer J. (2002) Knowledge seeking and location choice of foreign direct investment in the United States, *Management Science* **48**, 1534–1554. doi:10.1287/mnsc.48.12.1534.440

Chung W. and Kalnins A. (2001) Agglomeration effects and performance: a test of the Texas lodging industry, *Strategic Management Journal* **22**, 969–988. doi:10.1002/smj.178

Cohen W. M., Nelson R. R. and Walsh J. P. (2000) *Protecting their Intellectual Assets: Appropriability Conditions and Why US Manufacturing Firms Patent (or Not)*. National Bureau of Economic Research (NBER), Cambridge, MA.

Cohen W. M., Nelson R. R. and Walsh J. P. (2002) Links and impacts: the influence of public research on industrial R&D, *Management Science* **48**, 1–23. doi:10.1287/mnsc.48.1.1.14273

Cooke P. (2001) Regional innovation systems, clusters, and the knowledge economy, *Industrial and Corporate Change* **10**, 945–974. doi:10.1093/icc/10.4.945

De Faria P. and Sofka W. (2010) Knowledge protection strategies of multinational firms – a cross-country comparison, *Research Policy* **39**, 956–968. doi:10.1016/j.respol.2010.03.005

Defever F. (2006) Functional fragmentation and the location of multinational firms in the enlarged Europe, *Regional Science and Urban Economics* **36**, 658–677. doi:10.1016/j.regsciurbeco.2006.06.007

Del Barrio-Castro T. and García-Quevedo J. (2005) Effects of university research on the geography of innovation, *Regional Studies* **39**, 1217–1229. doi:10.1080/00343400500389992

DEYLE H. G. and GRUPP H. (2005) Commuters and the regional assignment of innovative activities: a methodological patent study of German districts, *Research Policy* **34**, 221–234. doi:10.1016/j.respol.2005.01.003

FEINBERG S. E. and GUPTA A. K. (2004) Knowledge spillovers and the assignment of R&D responsibilities to foreign subsidiaries, *Strategic Management Journal* **25**, 823–845. doi:10.1002/smj.396

FELDMAN M. (2003) The locational dynamics of the US biotech industry: knowledge externalities and the anchor hypothesis, *Industry and Innovation* **10**, 311–329. doi:10.1080/1366271032000141661

FLEMING L. and SORENSON O. (2004) Science as a map in technological search, *Strategic Management Journal* **25**, 909–928. doi:10.1002/smj.384

FRIEDMAN J., GERLOWSKI D. A. and SILBERMAN J. (1992) What attracts foreign multinational corporations? Evidence from branch plant location in the United States, *Journal of Regional Science* **32**, 403–418. doi:10.1111/j.1467-9787.1992.tb00197.x

FRITSCH M. and FRANKE G. (2004) Innovation, regional knowledge spillovers and R&D cooperation, *Research Policy* **33**, 245–255. doi:10.1016/S0048-7333(03)00123-9

FROST T. S., BIRKINSHAW J. M. and ENSIGN P. C. (2002) Centers of excellence in multinational corporations, *Strategic Management Journal* **23**, 997–1018. doi:10.1002/smj.273

GHEMAWAT P. (2001) Distance still matters, *Harvard Business Review* **79**, 137–147.

GHOSHAL S. and BARTLETT C. A. (1990) The multinational corporation as an interorganizational network, *Academy of Management Review* **15**, 603–625.

GOMES-CASSERES B., HAGEDOORN J. and JAFFE A. B. (2006) Do alliances promote knowledge flows?, *Journal of Financial Economics* **80**, 5–33. doi:10.1016/j.jfineco.2004.08.011

GRILICHES Z. (1998) Patent statistics as economic indicators: a survey. In GRILLICHES Z. (Ed.) *R&D and Productivity: The Econometric Evidence*, pp. 287–343. University of Chicago Press, Chicago, IL.

HALL B. H., JAFFE A. B. and TRAJTENBERG M. (2001) *The NBER Patent Citation Data File: Lessons, Insights and Methodological Tools.* National Bureau of Economic Research (NBER), Cambridge, MA.

HALL B. H., JAFFE A. B. and TRAJTENBERG M. (2003) *Market Value and Patent Citations. Revised Version of Market Value and Patent Citations: A First Look 1999.* NBER Working Paper. National Bureau of Economic Research (NBER), Cambridge, MA.

HALL B. H., JAFFE A. and TRAJTENBERG M. (2005) Market value and patent citations, *RAND Journal of Economics* **36**, 16–38.

HARHOFF D., HENKEL J. and VON HIPPEL E. (2003) Profiting from voluntary information spillovers: how users benefit by freely revealing their innovations, *Research Policy* **32**, 1753–1769. doi:10.1016/S0048-7333(03)00061-1

HEAD K., RIES J. and SWENSON D. (1995) Agglomeration benefits and location choice: evidence from Japanese manufacturing investments in the United States, *Journal of International Economics* **38**, 223–247. doi:10.1016/0022-1996(94)01351-R

HEJAZI W. and SAFARIAN A. E. (2002) Explaining Canada's changing FDI patterns. Unpublished manuscript, University of Toronto, Toronto, ON.

HILBER C. A. L. and VOICU I. (2010) Agglomeration economies and the location of foreign direct investment: empirical evidence from Romania, *Regional Studies* **44**, 355–371. doi:10.1080/00343400902783230

JAFFE A. B., TRAJTENBERG M. and HENDERSON R. (1993) Geographic localization of knowledge spillovers as evidenced by patent citations, *Quarterly Journal of Economics* **108**, 577–598. doi:10.2307/2118401

JAMES S. D. (2014) The use of voluntary public disclosure and patent strategies to capture value from product innovation, *Journal of Applied Business and Economics* **16**, 11–26.

JAMES S. and SHAVER J. (2014) *Motivations for Voluntary Public R&D Disclosures.* Working Paper. Fisher College of Business, Ohio State University, Columbus, OH.

KLOMP L. and VAN LEEUWEN G. (2001) Linking innovation and firm performance: a new approach, *International Journal of the Economics of Business* **8**, 343–364. doi:10.1080/13571510110079612

KUMAR N. (2001) Determinants of location of overseas R&D activity of multinational enterprises: the case of US and Japanese corporations, *Research Policy* **30**, 159–174. doi:10.1016/S0048-7333(99)00102-X

LANJOUW J. O. and SCHANKERMAN M. (2004) Patent quality and research productivity: measuring innovation with multiple indicators, *Economic Journal* **114**, 441–465. doi:10.1111/j.1468-0297.2004.00216.x

LEE K. and LIM C. (2001) Technological regimes, catching-up and leapfrogging: findings from the Korean industries, *Research Policy* **30**, 459–483. doi:10.1016/S0048-7333(00)00088-3

MANSFIELD E. (1995) Academic research underlying industrial innovations: sources, characteristics, and financing, *Review of Economics and Statistics* **77**, 55–65. doi:10.2307/2109992

MARSHALL A. (1920) *Principles of Economics: An Introductory Volume.* Macmillan, London.

MARTIN P. and OTTAVIANO G. I. P. (1999) Growing locations: Industry location in a model of endogenous growth, *European Economic Review* **43**, 281–302. doi:10.1016/S0014-2921(98)00031-2

MARX M. (2011) The firm strikes back non-compete agreements and the mobility of technical professionals, *American Sociological Review* **76**, 695–712. doi:10.1177/0003122411414822

MARX M., STRUMSKY D. and FLEMING L. (2009) Mobility, skills, and the Michigan non-compete experiment, *Management Science* **55**, 875–889. doi:10.1287/mnsc.1080.0985

MAURSETH P. B. and VERSPAGEN B. (2002) Knowledge spillovers in Europe: a patent citations analysis, *Scandinavian Journal of Economics* **104**, 531–545. doi:10.1111/1467-9442.00300

MCCANN P. and MUDAMBI R. (2004) The location behavior of the multinational enterprise: some analytical issues, *Growth and Change* **35**, 491–524. doi:10.1111/j.1468-2257.2004.00259.x

MCCANN P. and MUDAMBI R. (2005) Analytical differences in the economics of geography: the case of the multinational firm, *Environment and Planning A* **37**, 1857–1876. doi:10.1068/a37311

McFadden D. (1974) Conditional logit analysis of qualitative choice behavior, in Zarembka P. (Ed.) *Frontiers in Econometrics*, pp. 105–142. Academic Press, New York, NY.

McFadden D. and Train K. (2000) Mixed MNL models for discrete response, *Journal of Applied Econometrics* **15**, 447–470. doi:10.1002/1099-1255(200009/10)15:5<447::AID-JAE570>3.0.CO;2-1

Organisation for Economic Co-operation and Development (OECD) (2007) *Intellectual Assets and International Investment: A Stocktaking of the Evidence*. Report to the OECD Investment Committee DAF/INV/WD (2007) 6. OECD, Paris.

Piergiovanni R. and Santarelli E. (2001) Patents and the geographic localization of R&D spillovers in French manufacturing, *Regional Studies* **35**, 697–702. doi:10.1080/00343400120101434

Porter M. (2000) Locations, clusters, and company strategy, in Clark G., Feldman M. and Gertler M. (Eds) *The Oxford Handbook of Economic Geography*, pp. 253–274. Oxford University Press, Oxford.

Revelt D. and Train K. (1998) Mixed logit with repeated choices: households' choices of appliance efficiency level, *Review of Economics and Statistics* **80**, 647–657. doi:10.1162/003465398557735

Rosenkopf L. and Almeida P. (2003) Overcoming local search through alliances and mobility, *Management Science* **49**, 751–766. doi:10.1287/mnsc.49.6.751.16026

Rosenthal S. S. and Strange W. C. (2003) Geography, industrial organization, and agglomeration, *Review of Economics and Statistics* **85**, 377–393. doi:10.1162/003465303765299882

Schmoch U., Laville F., Patel P. and Frietsch R. (2003) *Linking Technology Areas to Industrial Sectors*. Final Report to the European Commission, DG Research.

Shaver J. M. and Flyer F. (2000) Agglomeration economies, firm heterogeneity, and foreign direct investment in the United States, *Strategic Management Journal* **21**, 1175–1193. doi:10.1002/1097-0266(200012)21:12<1175::AID-SMJ139>3.0.CO;2-Q

Shimizutani S. and Todo Y. (2008) What determines overseas R&D activities? The case of Japanese multinational firms, *Research Policy* **37**, 530–544. doi:10.1016/j.respol.2007.11.010

Song J., Almeida P. and Wu G. (2003) Learning-by-hiring: when is mobility more likely to facilitate interfirm knowledge transfer?, *Management Science* **49**, 351–365. doi:10.1287/mnsc.49.4.351.14429

Storper M. (1995) Regional technology coalitions an essential dimension of national technology policy, *Research Policy* **24**, 895–911. doi:10.1016/0048-7333(94)00810-8

Tappeiner G., Hauser C. and Walde J. (2008) Regional knowledge spillovers: fact or artifact?, *Research Policy* **37**, 861–874. doi:10.1016/j.respol.2007.07.013

Teece D. J. (1986) Profiting from technological innovation: implications for integration, collaboration, licensing and public policy, *Research Policy* **15**, 285–305. doi:10.1016/0048-7333(86)90027-2

Tether B. S. (2002) Who co-operates for innovation, and why: an empirical analysis, *Research Policy* **31**, 947–967. doi:10.1016/S0048-7333(01)00172-X

Thursby J. and Thursby M. (2009) *Here or There? A Survey of Factors in Multinational R&D Location: Report to the Government/University/Industry Research Roundtable (December 1, 2006)*. Kauffman Foundation Large Research Projects Research, Kansas City, MO.

Train K. (2003) *Discrete Choice Methods with Simulation*. Cambridge University Press, Cambridge.

Trajtenberg M. (1990) A penny for your quotes: patent citations and the value of innovations, *Rand Journal of Economics* **21**, 172–187. doi:10.2307/2555502

Trajtenberg M., Henderson R. and Jaffe A. (1997) University versus corporate patents: a window on the basicness of invention, *Economics of Innovation and New Technology* **5**, 19–50. doi:10.1080/10438599700000006

Warda J. (2006) *Tax Treatment of Business Investments in Intellectual Assets: An International Comparison*. OECD Science, Technology and Industry Working Papers. Organisation for Economic Co-operation and Development (OECD), Paris.

Zhao M. (2006) Conducting R&D in countries with weak intellectual property rights protection, *Management Science* **52**, 1185–1199. doi:10.1287/mnsc.1060.0516

Zhao M. and Alcácer J. (2007) *Global Competitors as Next-Door Neighbors: Competition and Geographic Concentration in the Semiconductor Industry*. Ross School of Business Paper, Ann Arbor, MI.

Industry-Specific Firm Growth and Agglomeration

MATTHIAS DUSCHL*, TOBIAS SCHOLL†‡, THOMAS BRENNER*, DENNIS LUXEN§ and FALK RASCHKE†‡

*Philipps University of Marburg, Department of Geography, Marburg, Germany

†House of Logistics & Mobility (HOLM) GmbH, Gateway Gardens, Germany

‡Philipps University of Marburg, Department of Geography, Frankfurt, Germany

§Karlsruhe Institute of Technology, Department of Informatics, Institute of Theoretical Informatics, Karlsruhe, Germany.

DUSCHL M., SCHOLL T., BRENNER T., LUXEN D. and RASCHKE F. Industry-specific firm growth and agglomeration, *Regional Studies*. This paper analyses the industry-specific relationship between industrial clustering and firm growth. Micro-geographically defined agglomeration measures based on travel-time distances and a flexible log-logistic decay function framework are used to study the spatial impacts of related economic and knowledge-generating activities in 23 industries. It is found that firms' growth prospects tend to be generally hampered by the agglomeration of own-industry employment, whereas the impact of proximate scientific activities systematically depends on the kind and age of industry. Furthermore, the optimal specifications of decay function that measures agglomeration effects considerably vary between both the industrsssies and the variables.

DUSCHL M., SCHOLL T., BRENNER T., LUXEN D. and RASCHKE F. 特定产业的企业成长与聚集，*区域研究*。本文分析特定产业在产业集群与企业成长之间的关联性。本研究运用根据旅行时间距离在微观地理层次所定义的聚集方法和弹性的对数罗吉斯衰减函数架构，研究二十三个产业中，相关经济与知识所引发的活动。研究发现，一般而言，企业的成长前景倾向受到同产业雇用的聚集所妨碍，而最近科学活动的影响则系统性地取决于产业的类别与年龄。再者，测量聚集效应的衰减函数的理想规格，在产业与变项之间皆有大幅的差异。

DUSCHL M., SCHOLL T., BRENNER T., LUXEN D. et RASCHKE F. La croissance des entreprises spécifiques à une industrie et l'agglomération, *Regional Studies*. Cet article analyse les relations spécifiques à une industrie entre le regroupement industriel et la croissance des entreprises. Pour étudier les effets spatiaux des activités économiques et de génération des connaissances auprès de 23 industries, on emploie des mesures d'agglomération définies d'un point de vue micro-géographique fondées sur la notion de distance exprimée en termes du temps de déplacement et sur le cadre log-logistique flexible d'une fonction de décroissance. Il s'avère que les perspectives de croissance ont généralement tendance à être entravées par l'agglomération de l'emploi dans les industries autochtones, alors que l'impact des activités scientifiques à proximité dépend systématiquement de la classification et de l'âge d'industrie. Qui plus est, les spécifications optimales de la fonction de décroissance qui mesure les effets d'agglomération varient sensiblement suivant l'industrie et en fonction des variables.

DUSCHL M., SCHOLL T., BRENNER T., LUXEN D. und RASCHKE F. Industriespezifisches Firmenwachstum und Agglomerationen, *Regional Studies*. In diesem Artikel wird der industriespezifische Zusammenhang zwischen industriellen Clustern und Firmenwachstum analysiert. Zur Untersuchung der räumlichen Effekte von verwandten ökonomischen und wissenserzeugenden Aktivitäten werden mikrogeographisch definierte Agglomerationsmaße, basierend auf Fahrtzeiten und einer flexiblen log-logistischen Distanzfunktion, verwendet. Die Ergebnisse für 23 Industrien zeigen, dass die Wachstumsaussichten von Firmen im Allgemeinen durch die Agglomeration von Beschäftigten aus der gleichen Industrie beeinträchtigt werden, während der Einfluss benachbarter wissenschaftlicher Aktivitäten systematisch von der Art und dem Alter der Industrie abhängt. Darüber hinaus variiert die optimale Spezifikation der Distanzfunktionen, welche Agglomerationseffekte messen, deutlich zwischen den Industrien und Variablen.

DUSCHL M., SCHOLL T., BRENNER T., LUXEN D. y RASCHKE F. Crecimiento de las empresas y aglomeración específicos para las industrias, *Regional Studies*. En este artículo analizamos la relación específica de la industria entre la aglomeración industrial y el crecimiento de las empresas. Estudiamos los impactos espaciales de actividades relacionadas económicas y actividades que generan conocimiento en 23 industrias a partir de medidas de aglomeración definidas micro-geográficamente basadas en las distancias según el tiempo de viaje y un marco flexible de función de decrecimiento según un modelo log-logístico. Observamos que las perspectivas de crecimiento de las empresas tienden a estar generalmente obstaculizadas por la aglomeración del empleo en la propia industria, mientras que el impacto de las actividades científicas cercanas depende sistemáticamente del tipo y la edad de la industria. Asimismo las especificaciones óptimas de la función de decrecimiento que mide los efectos de aglomeración varían considerablemente entre las industrias y las variables.

INTRODUCTION

Geographical location has for a long time been 'a neglected determinant of firm growth' (AUDRETSCH and DOHSE, 2007), but recently an increasing bulk of the literature examines the impact of being located within agglomerations and industrial clusters or in proximity to universities on the performance of firms. However, empirical findings are still contradictory. This is not surprising, as one of the few invariables of industrial dynamics is the heterogeneity of firms (DOSI *et al.*, 2010). From the existing literature, four key issues are known as a potential source of contradicting results. First, there are strong differences between industries. For instance, agglomeration economies differ between manufacturing and service industries as well as at finer levels of disaggregation (e.g., BEAUDRY and SWANN, 2009). Second, processes and mechanisms differ with the industry's age and particularly its stage in the industrial life cycle. Empirical investigations suggest that agglomeration of economic and knowledge-generating activities is more important in the initial phase of an industry (e.g., POTTER and WATTS, 2011). Third, agglomeration can be measured with different statistical methods. Fourth, the spatial dimension used within the methods matters. Results might depend on the chosen regional level, i.e. if the investigation is done on the city level, zip codes or other functional definitions of regions. Some investigations even report contradicting results using the same dataset when changing from one aggregation level to another (e.g., BUERGER *et al.*, 2010).

Recently, much effort has been undertaken to improve the understanding of spatial dependencies by shifting from aggregated large-scale investigations to more micro-geographic data-driven approaches (e.g., DURANTON and OVERMAN, 2005). Beside the issue of availability, micro-geographic data also require new techniques in order to integrate them into econometric models. Given the few existing publications that deal with micro-geographic data, the literature remains unclear on how the methodological challenges can be met. While micro-geographic data enhance the validity of research findings on spatial matters of firm growth, investigations have yet mostly neglected the physical nature of a firm's geographic location. As regards the access to growth-relevant sources of agglomeration economies, its location relative to the road network plays an important role.

Given these yet unresolved questions, the aim of this paper is to re-examine the effects of external factors to the growth prospects of incumbent firms. The literature provides clear evidence for the fact that agglomerations or the proximity of universities impact on firm growth, but the details of these impacts are not well studied so far. More detailed empirical knowledge about these impacts will provide hints about the underlying mechanisms. The theoretical literature puts forward various potential mechanisms, but little can be said about their (relative) relevance. This paper goes into detail in two aspects. First, it studies several industries separately in order to examine the role of industrial characteristics, such as maturity. Second, it takes a detailed look at spatial aspects in order to examine the spatial range of relationships, which provides hints about the mechanisms behind the detected relationships. This is achieved by geo-locating firms into space and by calculating travel-time distances to all related economic and knowledge-generating activities. Firms' access to these activities is spatially discounted by a flexible log-logistic distance-decay function, which can be deduced from behavioural assumptions and which is further specified based on empirical data. Using a quantile regression framework, the impact of nearby economic and knowledge-generating activities on employment growth of German firms is compared across 23 industries, i.e. groups of related industries as defined by the European Union Cluster Observatory.

The findings suggest that being located in agglomerations of own-industry employment does not increase but rather reduces firms' growth prospects. In contrast hereto, the impact of being located in proximity to knowledge-generating activities depends more systematically on the type and age of the industry – in less mature, more knowledge-intensive industries the relationship with firm growth tends to be positive. Also, the spatial scale of what proximity means varies across industries. In order to account for the heterogeneity of the analysed industries, results from three representative cases are finally discussed in more depth.

The paper is structured as follows. In the second section, expectations on the industry-specific impacts of nearby economic and knowledge-generating activities on firm growth are deduced from the existing literature. Data issues are described in the third section; while the methodology of a data-driven distance-decay function specification and a quantile regression framework with spatially discounted variables is outlined in the fourth section. The fifth section presents and discusses the results. The sixth section concludes.

LITERATURE

The impact of agglomeration on firm growth

A rich body of academic work exists that can be labelled with the term 'agglomeration theory', among which the theory of industrial clusters (PORTER, 2000), which comprises being located in spatial proximity to similar firms and associated institutes like universities, has attracted much interest in both science and politics. The impact of agglomeration economies has been studied from many different points of views. Whilst the effect on start-ups (e.g., BUENSTORF and KLEPPER, 2009), firm survival (e.g., NEFFKE *et al.*, 2012), or innovative (e.g., FORNAHL *et al.*, 2011) and productive performance (RIGBY and ESSLETZBICHLER, 2002) is well documented, FRENKEN *et al.* (2011) identify in their survey on *industrial dynamics and economic geography* a gap in the empirical understanding of the effects on the growth performance of incumbent firms.

Porter and co-authors find in several papers that within clusters wages, innovativeness and entrepreneurship are higher (PORTER, 2003; DELGADO *et al.*, 2012). Benefiting from the co-location with other similar firms is one of the major arguments within cluster theory (PORTER, 2000). This implies that firms which are located within agglomerations of their industry should have higher growth potentials. The comprehensive study by BEAUDRY and SWANN (2009) on 56 two-digit industries in the UK supports this argument: in about half of the industries, being located within agglomerations of workers in the same sector has a positive impact on firms' growth prospects.

In contrast, other researchers do not find higher survival or growth rates for firms located in clusters (BUENSTORF and KLEPPER, 2009). While still acknowledging the positive influence of clusters on spin-offs, FRENKEN *et al.* (2011, p. 2) conclude that 'there is little evidence that clusters enhance firm growth and survival'. Various empirical investigations show that industrial clusters contribute to firm growth only under certain circumstances and that it is of high importance which constituent parts (agglomeration of similar firms, research institutions, etc.) are observed. However, further evidence is missing and various mechanisms are claimed to be responsible for this benefit. Therefore, a more distinguishing perspective might be helpful.

One explanation of why firms might benefit from industrial clusters is based on phenomena like innovation, learning and knowledge spillovers (MALMBERG *et al.*, 2000). Growth-relevant knowledge is generated by competing and cooperating firms, but also by research activities in universities or public research institutes. In the latter case, the literature is less ambiguous. Many studies show a positive link between the presence of universities and firms' innovation (e.g., JAFFE, 1989) and growth performance (e.g., AUDRETSCH and LEHMANN, 2005; CASSIA *et al.*, 2009; RASPE and VAN OORT, 2011). This relates to research on the relationship between regional knowledge intensity and firm performance, which (indirectly) assumes spatially bounded knowledge spillovers to be one of the main mechanisms of agglomeration economies (FRENKEN *et al.*, 2011). Again, an industry-specific perspective is suggested: knowledge spillovers from external research activities should matter most for firms of science-based or knowledge-intensive industries, which also can be expected to engage more in internal research activities as well as to provide the required absorptive capacities (KOO, 2005).

Besides the kind of industry, the industry's age and its life cycle stage can help to understand the multifaceted influences of industrial clusters on firm growth (FELDMAN, 1999). This is also reflected by the recent focus on the dynamics of agglomeration economies during industrial life cycles (e.g., NEFFKE *et al.*, 2011). In early life cycle phases, which often coincidence with the emergence of new industrial clusters, the rates of start-ups and spin-offs tend to be high. Local conditions, like the presence of related industries, public research institutes or universities, play a major role in the initial development of clusters. At more mature stages, however, market growth slows down and a kind of equilibrium is reached (BRENNER and SCHLUMP, 2011). Empirical evidence suggests that under these circumstances firms do not benefit any more from being located in industrial clusters (AUDRETSCH and FELDMAN, 1996), and their growth prospects might even be hampered due to increasingly prevailing negative agglomeration economies such as intensified competition (POTTER and WATTS, 2011).

To conclude, it is expected that industrial clusters have manifold effects on firm growth: effects of the agglomeration of related economic activities are ambiguous, depending on the kind of industry and its stage in the industry life cycle. In contrast, the effects of proximate knowledge-generating activities should be rather positive, especially so in less mature, more knowledge-intensive industries.

Spatial matters of firm growth and agglomeration

Although the impact of industrial clusters on firm growth is highlighted in many studies, the literature has remained quite silent on the spatial range of their influence and the actual definition of space. For most of the papers, the definition of space arises from the used dataset, i.e. the spatial aggregation level of the data such as cities or regions. For quantitative-driven investigations, mostly regression models, this spatial definition concerns both the dependent and the independent variable.

Table 1 provides a non-exhaustive overview of the different approaches that can be found in the literature. Starting with the dependent variable, two general concepts can be separated: The macro-approach investigates growth effects on a regional level while the micro-approach deals with growth effects on single firms. Empirical evidence exists that the clustering of industries exerts a positive impact on regional economic performance, both for the entire regional economy (e.g., DELGADO *et al.*, 2010) as well as at the disaggregate level of industries within the region (e.g., DELGADO *et al.*, 2012). This evidence is refined by the insight that the underlying mechanisms like knowledge spillovers are most pronounced in regions where at the same time the variety and the relatedness of the agglomerated industries are highest (FRENKEN *et al.*, 2007). Spatially more sophisticated approaches deal with neighbourhood effects – how neighbouring regions influence the growth of a specific region. Recently, however, both the growing access to firm data and to computational power have led to a focus on micro-approaches. Investigating growth at the firm level allows for a more sophisticated testing as firm-specific variables can be included and intra-regional heterogeneity can be observed. While the dependent variable of the micro-approach is the growth rate of a single firm and therefore not aggregated, most studies explain growth processes by means of characteristics of the region in which the firm is located. Here, the literature has brought forward manifold regional measures for concentration, diversity or competition, ranging from simple counts, relative measures like the location quotient (LQ) to more complex, derivative measures. Because imperfect competition and heterogeneous firms are defining characteristics of the economic landscape, regions as consistent and homogenous aggregates do not exist in reality. As a consequence, regionalization, an ex-post abstraction of the continuous landscape, would imply a huge loss of information (for an extensive discussion on this issue, see PINSKE and SLADE, 2010; or HARRIS, 2011). Thus, analogous to the macro-approaches, results are affected by the arbitrariness of

Table 1. Literature overview

Approach	Observation	Industry focus	Modifiable areal unit problem (MAUP)	Examples
Macro (regions)	Intra-region effects	No	Yes	FRENKEN *et al.* (2007)
		Comparison		GLAESER *et al.* (1992), HENDERSON *et al.* (1995), PORTER (2003), SPENCER *et al.* (2010)
	Neighbourhood effects	No		KUBIS *et al.* (2009), DELGADO *et al.* (2010), ARTIS *et al.* (2012)
		Comparison		DELGADO *et al.* (2012)
Micro (firms or plants)	Intra-region effects	No	Yes	GUISO and SCHIVARDI (2007), CAINELLI (2008), BOSCHMA *et al.* (2009), CASSIA *et al.* (2009), ANDERSSON and LÖÖF (2011)
		Comparison		RIGBY and ESSLETZBICHLER (2002), HENDERSON (2003), AUDRETSCH and DOHSE (2007), BEAUDRY and SWANN (2009), WENNBERG and LINDQVIST (2010)
	Distance bands	No	No	HOOGSTRA and VAN DIJK (2004), BALDWIN *et al.* (2008), ERIKSSON (2011)
		Comparison		ROSENTHAL and STRANGE (2003), BALDWIN *et al.* (2010), GRAHAM (2009)
	Distance decay	No		AUDRETSCH and LEHMANN (2005), LYCHAGIN *et al.* (2010)
		Comparison		VAN SOEST *et al.* (2006), GRAHAM *et al.* (2010), DRUCKER and FESER (2012), DRUCKER (2012)

regional boundaries and moreover by the chosen level of aggregation. This issue of zoning and scaling was first described by OPENSHAW (1984) and coined with the term 'modifiable areal unit problem' (MAUP). By varying the spatial scale of analysis, BUERGER *et al.* (2010) as well as WENNBERG and LINDQVIST (2010) show empirically that the MAUP is highly relevant for agglomeration economies.

Avoiding the MAUP requires two methodological aspects. First, the aggregation level of the data should be as low as possible. DURANTON and OVERMAN (2005) refer to this as micro-geographic data, which were obtained in the present case by computing the easting and northing of each firm's municipality. Municipalities represent the lowest aggregation level in Germany, currently with a number of 11 249 and an average size of 31.6 km^2. Second, distance-based methods have to be applied for the calculation of the independent variables. One method is using distance bands, i.e. counting the observance of firms at specific radii (e.g., ROSENTHAL and STRANGE, 2003). Another approach is the use of distance-decay functions that build proxy values of agglomeration by summing up localizable activities multiplied by inverted distances. Various specifications of both the distance bands and the decay functions exist in the literature. Concerning the latter approach, mostly simple linear (e.g., AUDRETSCH and LEHMAN, 2005) or exponential decay functions (e.g., DRUCKER and FESER, 2012) are used, although DE VRIES *et al.* (2009) have shown that a log-logistic function is best suited for modelling spatial interactions, in their case, the effect of transportation costs on commuting flows. This function, derivable from behavioural assumptions, represents a rather flexible approach, to which the exponential decay and even the distance bands are only special cases. Because agglomeration economies are reported for a wide range of different distances, from a narrow local to supra-regional scale, in the present approach the best-fitting decay function will be identified based on empirical data and for each industry separately.

Beside the choice of the distance model, results may also depend on the way how distance between firms is computed. The vast majority of distance-based investigations uses orthodromic distances (e.g., kilometres or miles), although this might cause errors if the 'economic' distance between firms deviates from the orthodromic distance, for instance, if firms are located in mountainous or less well-connected regions (DURANTON and OVERMAN, 2005). Obviously, driving distance or travel time are more appropriate, given that agglomeration economies are assumed to arise from low transportation costs or the convenience of face-to-face contacts. One of the few exceptions using travel times is the work of AUDRETSCH and LEHMANN (2005), where the growth of firms is investigated with respect to the firms' driving distance to their closest university. However, studies where driving distances are computed to thousands of locations are, due to the high computational costs of route planning, very rare. Using an efficient many-to-many route planning algorithm, introduced by KNOPP *et al.* (2007), travel times are computed here between all German municipalities, allowing agglomeration effects on firm growth to be investigated from a more realistic spatial perspective.

With respect to the discussed literature, the paper at hand belongs to the group of micro-approaches as it observes the growth rates of each individual firm. It uses a flexible distance-decay function and compares agglomeration effects resulting from related economic and scientific knowledge-generating activities across disaggregated industries. From a methodological point of view, this paper differs from the existing literature regarding two aspects. First, instead of spherical distances, travel time in minutes is used. Secondly, a specific distance-decay function is not anticipated but its identification is included in the analysis in order to detect possible differences among industries.

DATA

Definition of industries

The 23 industries used for the current analyses were taken from the European Union Cluster Observatory and can be seen as a standard definition for industry-related policy programmes on regional development in Europe. The definition goes back to a US cluster mapping project undertaken by Porter in the early 2000s and is based on the distinction between local and natural-resource-driven industries, on the one hand, and export-oriented traded industries, on the other (PORTER, 2003). The latter industries were grouped at the four-digit level of the standard industrial classification (Nomenclature statistique des activités économiques dans la Communauté européenne (NACE) Rev. 2) according to co-location patterns within economic activities from data across the United States and led to groups of related industries. Porter's analysis concludes that the regional presence of those activities that tend to cluster in space can be seen as a driver for regional economic performance and the positive development of embedded firms (PORTER, 2003; WENNBERG and LINDQVIST, 2010). Therefore, the definition of industries as groups of co-located and, hence, related activities is suitable to survey the relationship between industrial clustering in space and firm growth.

Dependent variable

The BvD Amadeus database discloses the address of the firms' headquarters' location. Operational and strategic decisions are often made within this organizational unit. Although research and development (R&D)-intensive foreign direct investments have recently become more important (UNITED NATIONS

CONFERENCE ON TRADE AND DEVELOPMENT (UNCTAD), 2005), many of the firms' R&D activities still remain located close to the headquarters (GUIMÓN, 2009). Even for multinational enterprises, a home bias for innovation activities is evident (COHEN *et al.*, 2009; DUNNING and LUNDAN, 2009). Therefore, the work of BEAUDRY and SWANN (2009) is followed by assuming that it is the regional environments of a firm's headquarters that is most decisive in affecting the growth prospects and the decision of the organization on taking up employees. This rationale breaks down for larger firms, which tend to be less focused on their headquarters, but disperse activities in many increasingly independent establishments across the country and even beyond. Therefore, the analysis is restricted to firms with no more than an annual average of 1000 employees.[1] Also, very small firms with fewer than five employees, which growth processes are known to be rather erratic, are excluded (COAD, 2009).

Growth rates are calculated by taking the difference of the natural logarithms of the size S (measured by the entire stock of employees) of firm i between two successive years t:

$$g_{i,t} = \log(S_{i,t+1}) - \log(S_{i,t}) \tag{1}$$

Confronted with an unbalanced panel from 2004 to 2010, yearly growth rates are pooled together.[2] In the course of one year, firms essentially face three options: they may expand, shrink or remain at the previous level. Zero-growth events, usually quite abundant for employment, make up 44.5% of the original data. A considerable but unknown share of these events can be attributed to a lack of regular updating of database entries, which are simply extrapolated from previous years. Including these events would bias the assessment of the impact of agglomeration on firm growth. Besides this data issue, firms that opt not to grow might be distinct from actually changing (expanding or shrinking) firms. Although being an economically rational choice in the absence of any changes in business opportunities, this option is often preferred even in cases when opportunities have changed. To name just a few examples, firms might be reluctant to expand because the inclusion of new employees is costly as it implies reorganization of internal tasks and management functions, or the fear of losing control might frighten some managers (COAD, 2009). In a similar vein, firms can be reluctant to shrink despite reduced business opportunities. Firms invest in building up redundancies in difficult times instead of immediately dissolving existing working contracts, or managers might be not fully aware of the necessary down-sizing. Because from a technical point of view it is impossible to distinguish between data problems and the various other reasons why firms do not grow, only growth events in which the size of firms actually changes are analysed here.

Industries with fewer than 1000 yearly growth events are omitted due to robustness issues. The remaining 23 industries are listed in Table 2, together with their number of pooled growth events $g_{i,t}$, number of firms and the average age of these firms.

Table 2. Overview of the analysed industries

ID	Name	N ($g_{i,t}$)	N (firms)	Age (years)
1	Agricultural products	1077	688	19
2	Automotive	1721	632	20
3	Building fixtures and equipment	2529	1115	23
4	Business services	5057	2417	13
5	Chemical products	1385	504	24
6	Construction	6278	3057	24
7	Distribution	5108	2488	22
8	Entertainment	1179	558	14
9	Financial services	1108	540	14
10	Heavy machinery	1041	391	21
11	Instruments	1336	539	25
12	Information technology (IT)	2668	1161	15
13	Media and publishing	3038	1654	23
14	Medical devices	1068	573	19
15	Metal manufacturing	7189	3265	25
16	Paper products	2159	970	24
17	Plastics	1861	748	24
18	Processed food	4652	2222	26
19	Production technology	5273	2072	24
20	Telecom	1406	524	21
21	Textiles	1097	488	26
22	Tourism and hospitality	2100	1472	18
23	Transportation and logistics	2406	1013	17

Independent variables

Firms' potential to benefit from industrial clusters is specific to characteristics of the firms as well as of the corresponding regions (BEUGELSDIJK, 2007; ERIKSSON, 2011). Therefore, the independent variables consist of three different kinds. First, relevant demographic properties of the firms are controlled for. Second, measures of the general environment of the region in which the firm is located are included. Third, the focus of this paper lies in firm-specific location variables reflecting agglomerations of related economic and scientific activities.

Control variables: demographic and regional variables. Building upon the literature on firm growth, which mostly extends a Gibrat-like growth regression (for an overview, see COAD, 2009), the logarithm of size, age and whether or not it is a subsidiary firm are controlled for. It counts as a stylized fact in industrial dynamics that firm growth is negatively related to both size and age. In addition, two variables are chosen to control for the general regional environment. Urbanization economies per se, which are rather independent from the surrounding industrial structure (BUERGER *et al.*, 2012) and which might be both positive or negative, can be measured by the population density of the corresponding district in which a firm is located (ERIKSSON, 2011).[3]

The unemployment rate of the firm's regional labour market (as defined by ECKEY *et al.*, 2006) reflects the vitality of the regions' socio-economic conditions. In the special case of Germany it also accounts for structural differences along the east–west and north–south divide. Data for both variables are obtained from the German Federal Statistical Office. The global macroeconomic recession of 2008–10 systematically lowered the firms' growth prospects. Therefore, a dummy variable for the crisis years is constructed. Finally, the absolute location within Germany might influence the magnitude of agglomeration economies. Potential cross-border effects cannot be considered for discriminating firms located close to the border. Due to historical reasons, two dummies are constructed: one for the location in border regions with the new member states of the European Union and one for all other border regions.

Firm-specific location variables: own-industry employment and publications. In contrast to the regional control variables that account for a rather diffuse socio-economic environment (or 'social filter', as denominated by RODRÍGUEZ-POSE and CRESCENZI, 2008), other economic and knowledge-generating activities can be traced back to concrete localizations in space: firms compete, cooperate and learn from each other, and new scientific knowledge originates from universities and research institutes. These related economic and knowledge-generating activities can be approximated by the number of employees in the same group of related activities (hereafter, industry) and scientific publications, respectively. The Federal Employment Agency provides data on industry-specific employment for municipalities, the lowest aggregation level in Germany. Data on scientific publications were collected from the ISI Web of Science and assigned to municipalities on the basis of authors' addresses. Since the two variables will be the focus of this paper, the subsequent section explains in more detail how firm-specific location variables can be constructed by discounting these geo-localized activities with their distances to the firm's location. Therefore, bilateral travel times are calculated by exploiting results from graph theory and data on the German road network from the OpenStreetMap project; the algorithms are described in DUSCHL *et al.* (2014) and more extensively by GEISBERGER *et al.* (2010). Intra-municipality distances are set to 5.01 min, the average bilateral travel time between 1000 randomly drawn pairs of firms' address locations, each belonging to the same municipality.

MODEL AND ESTIMATION

Construction of firm-specific location variables

Employees in the same industry and scientific publications are discounted by an industry-specific distance-decay function $f(d_{im})$ based on travel time distances d_{im} between the places of firms i and the municipalities m. The firm-specific variables for agglomeration of related economic ($AGGL$) and knowledge-generating activities ($KNOW$), after normalizing with:

$$\mu_{\mathrm{AGGL}} = \frac{\sum_m empl_{m,t}}{\sum_i \sum_m f(d_{im}) empl_{m,t}}$$

$$\mu_{\mathrm{KNOW}} = \frac{\sum_m publ_{m,t}}{\sum_i \sum_m f(d_{im}) publ_{m,t}}$$

read:

$$\begin{aligned} AGGL_{i,t} &= \mu_{\mathrm{AGGL}} \sum_m f(d_{im}) empl_{m,t} \\ KNOW_{i,t} &= \mu_{\mathrm{KNOW}} \sum_m f(d_{im}) publ_{m,t} \end{aligned} \tag{2}$$

These spatially discounted variables can be included in a simple linear model (for spatial econometric issues, see ANDERSSON and GRASJÖ, 2009):[4]

$$\begin{aligned} g_{i,t} = \alpha + \beta_1 AGGL_{i,t} + \beta_2 KNOW_{i,t} \\ + \sum_{j=3}^{10} \beta_j x_{i,t,j} + \varepsilon_{i,t} \end{aligned} \tag{3}$$

where α and β represent the coefficients to be estimated; and x is the seven firm- and region-specific control variables. The error term is denoted by $\varepsilon_{i,t}$. The applied normalization procedure allows for an interpretation of the corresponding regression coefficients as the impact of one additional employee or publication on the growth of a firm with a given distance. After briefly introducing quantile regression techniques as an adequate estimation method in the context of firm growth, the still outstanding specification of the distance-decay function $f(d)$ will be discussed in the following sections.

Estimation using quantile regression techniques

It is one of the stylized facts of industrial dynamics that firm growth rates are not normally distributed, but show fat tails (for an overview on empirical studies for different countries, see COAD, 2014). Therefore, quantile regression techniques, which are robust to outliers in the dependent variable and free from any distributional assumption in the error term (BUCHINSKY, 1998), are more appropriate. Besides, the specific conditional quantiles of strongly expanding ($\theta_{0.75}$) and declining ($\theta_{0.25}$) firms can be analysed in addition to the median growing firm ($\theta_{0.5}$). The intuition is that high growth events, a dominant feature of firm growth, rely differently on internal as well as external factors. Technical details are described in KOENKER (2005). This section only points out that, similar to ordinary least squares (OLS) regression, the coefficient estimates can be interpreted as partial derivatives, meaning the impact of a 1 unit change of an independent variable on the firms' growth rate at the θth quantile holding all other variables fixed. Standard errors are obtained by bootstrapping based on 200 replications to alleviate the problem of underestimation in the presence of heteroskedastic error distributions (e.g., GOULD, 1992).

Identification of decay function parameters

Social interactions are fundamental to all mechanisms that underlie agglomeration economies, like labour market pooling, contracting with suppliers and customers, transfer of knowledge, but even local competition. From simple transaction cost reasoning, the frequency of interactions should decay with distance. Moreover, the literature on commuting behaviour (JOHANSSON *et al.*, 2003; ANDERSSON and KARLSSON, 2007) shows that the negative travel time sensitivity is not linear in space, but varies between different geographical scales: within a narrow local context, interactions can take place at short notice and are primarily governed by randomness (THORSEN *et al.*, 1999). Thus, within agglomerations interactions are only marginally affected by distance. At some threshold distance, however, the minimal cost principle predominates and, consequently, the frequency and contribution of growth-relevant economic interactions become highly distance sensitive and may decrease rapidly. This threshold can be said to define the range of the region from a firm's perspective. For very long distances, geography ceases to matter once again. Mathematically, these behavioural assumptions can be expressed as an 'S'-shaped and downward sloping log-logistic decay function of travel time d:

$$\begin{aligned} f_{r,s}(d) &= 1/(1 + r^{-s} * \exp(s * \log(d))) \\ &= 1/(1 + (d/r)^{-s}) \end{aligned} \quad (4)$$

where r and s represent two parameters that describe the shape of the curve (for technical details, see DE VRIES *et al.*, 2009). Parameter r determines the location of the curve's inflection point; and parameter s its degree of steepness. The curve starts rather flat with the value of 1, becomes steeper and then gradually flattens again as it approaches 0. If s becomes 1, the curve takes the shape of a negative exponential function. If s tends towards infinity, the function resembles a binary distance circle, with values of 1 for distances below r, and 0 for distances above r. Keeping r constantly at 90 min, Fig. 1 depicts five curves for different values of s.

To identify the best specification of $f_{r,s}(d)$, the two variables *AGGL* and *KNOW* are regressed on the growth rates for each industry, for each quantile θ, as well as for each possible integer combination of the parameters r and s within the intervals [5, 300] and [1, 20], respectively. This means r is allowed to vary between the minimal distance of 5 min, which corresponds to the intra-municipality distances, and 300 min, which is half the maximal travel distance found in Germany. The lower bound for s is mathematically given by 1, whereas the upper bound is set to 20, in which case the properties of a binary distance circle are approached. The smallest log-likelihood value gives the best fitting combination of r and s. Furthermore, the confidence intervals around these parameters, which contain plausible alternative specifications of $f_{r,s}(d)$, are determined using the likelihood ratio test. Besides a straightforward single optimum scenario like in production technology for the variable *KNOW* at $\theta_{0.5}$, multiple optima are also

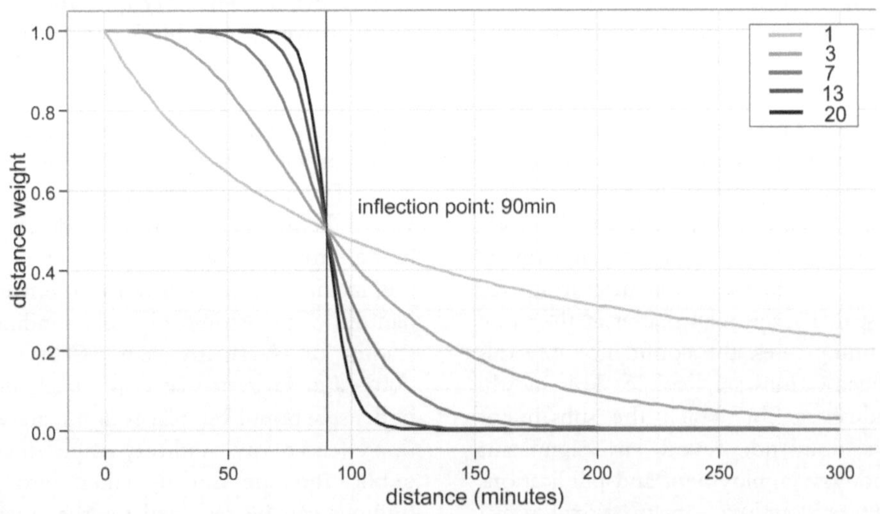

Fig. 1. Log-logistic decay function for inflection point r = 90 min and different degrees of steepness, s

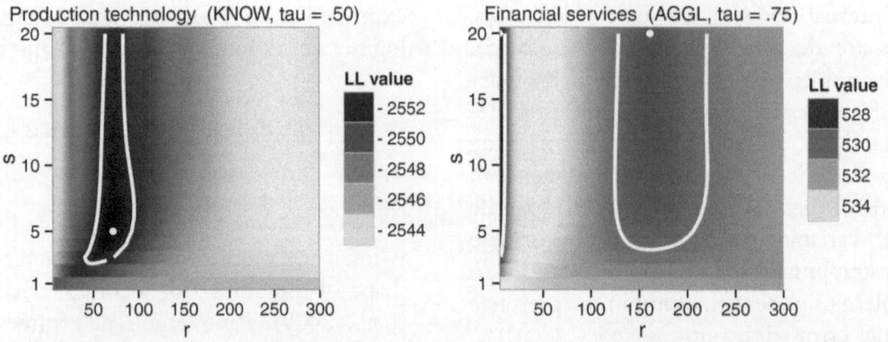

Fig. 2. Examples of the decay parameters optimization procedure with one minimum (left) and two minima scenarios (right). Minima are shown as white points, confidence interval borders are shown as white curves

feasible (Fig. 2): firm growth in financial services correlate with related employees at a narrow local scale, expressed by a decay function with r = 5 min and a sharply declining shape of s = 20, and simultaneously at a wider spatial scale with r = 161 and s = 20. In cases such as the latter, all significantly distinguishable optima will be included into the regression model as separate decay function specifications. However, never more than two optima are identified in any single case.

RESULTS

Distance-decay function specification

Fig. 3 provides an overview on the estimated parameters of the distance-decay function for each industry. The optimization procedure tends to converge to the following decay function specifications: in most cases, the specifications are rather similar to distance bands, with s = 20, and sometimes even significantly outperform the exponential decay function, which is often used in the literature. Here, the effects may abruptly decay at very short distances of a few minutes, which indicate the predominance of municipality-level and intra-regional effects, like in processed food (for *AGGL*) or heavy machinery (for *PUBL*), or even at distances beyond traditionally defined regional boundaries, like in medical devices or textiles, for which publications within the driving distance of 2 or 3 hours matter. In other cases, the impact of the location-specific variables on firm growth does not abruptly cease at any outstanding boundary, but decays exponentially, like in tourism and hospitality (for *AGGL*) or production technology (for *PUBL*). Furthermore, in some industries, like financial services (for *AGGL*) or metal manufacturing (for *PUBL*), different spatial scales may matter at the same time. In the remaining cases, the confidence intervals suggest that all specifications are feasible. This mostly holds true in industries, for which the subsequent regression analysis does not reveal any significant impact of own-industry employment and publications on firm growth. Here, cognitive, organizational, social or institutional proximity might be more relevant (BOSCHMA, 2005). To conclude, the apparent industry-specific heterogeneity highlights the importance of a sound and flexible distance-decay function specification to assess the impact of industrial clusters, which otherwise would be biased by the MAUP.

Control variables

The estimated coefficients of the control variables are in line with the current literature on firm growth and industrial dynamics. For the sake of brevity, the main findings are only summarized.[5]

First, yearly growth rates always negatively correlate with the firms' size and in most industries also with age. This confirms the literature (e.g., EVANS, 1987) rejecting Gibrat's law that assumes that growth is independent of size (and age). The revealed relationships become even more pronounced for highly growing firms at the conditional quantile $\theta_{0.75}$. Only for shrinking firms, at $\theta_{0.25}$, the growth relationship vanishes for age and becomes even positively significant for size in most industries. These patterns simply imply that the growth of larger and older firms is less volatile: they are less likely to realize large growth jumps, and at the same time they are less prone to strong negative growth events. Since HYMER and PASHIGIAN (1962), this negative relationship between growth rate variance and firm size is well studied.

Second, in one-third of the analysed industries, population density, a general measure of urbanization economies comes along with lower average growth rates. The only exception is found in entertainment, where being located in high-density districts means higher growth prospects. At the lower and higher quantile, its influence diminishes and remains significant in a handful of industries only. This finding confirms other studies on Germany (e.g., OTTO and FORNAHL, 2008) and suggests that cost aspects due to congestion in densely populated places dominate when agglomeration effects of own-industry employment and proximate publications are directly taken into account. Similar findings can be reported for the unemployment rate, measuring the general structural and economic

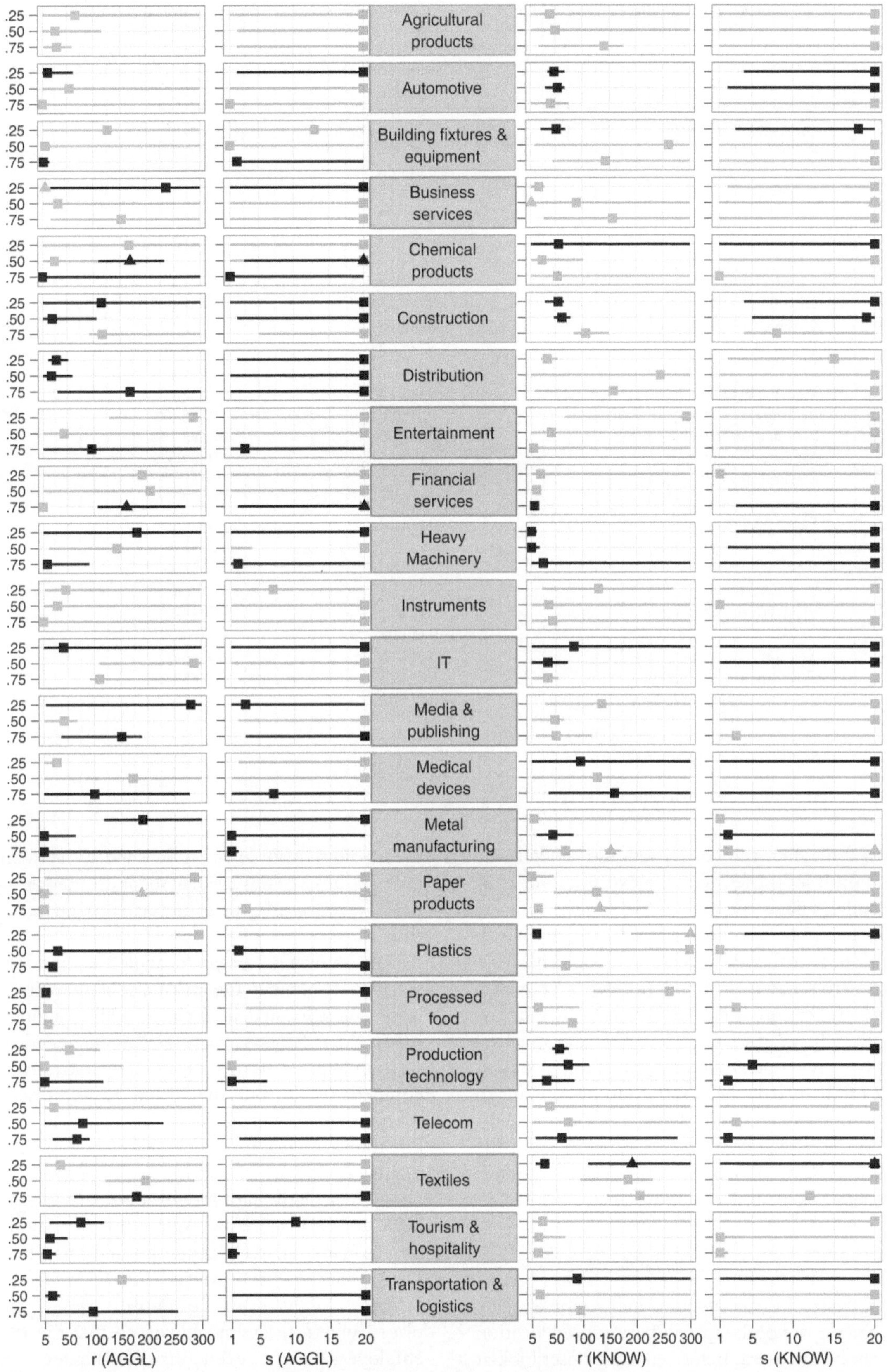

Fig. 3. Estimated distance-decay function parameters r (inflection point) and s (degree of steepness) with confidence intervals. Significant results (at 5%) are coloured black; a possible second optimum is indicated by a triangle

Table 3. Coefficients for agglomeration of related economic (AGGL) at different quantiles

ID	Name	*AGGL*		
		$\theta_{0.25}$	$\theta_{0.5}$	$\theta_{0.75}$
1	Agricultural products	–0.00015	–0.00004	–0.00002
2	Automotive	–0.00001*	0.00000	0.00004
3	Building fixtures and equipment	0.00006	–0.00003	–0.00011**
4	Business services	–0.00015*	–0.00004	–0.00011
		–0.00003		
5	Chemical products	–0.00004	–0.00000	–0.00013**
			–0.00010*	
6	Construction	0.00025***	0.00005*	0.00008
7	Distribution	0.00014*	0.00007*	0.00069*
8	Entertainment	0.00022	–0.00014	–0.00037*
9	Financial services	0.00002	–0.00002	–0.00001
				–0.00004*
10	Heavy machinery	–0.00013*	0.00005	–0.00014*
11	Instruments	0.00008	0.00007	–0.00002
12	Information technology (IT)	–0.00026*	–0.00020′	0.00014′
13	Media and publishing	0.00033*	–0.00004	0.00026*
14	Medical devices	–0.00003	–0.00021	–0.00029*
15	Metal manufacturing	–0.00013**	–0.00007*	–0.00009*
16	Paper products	0.00012	0.00000	–0.00001
			–0.0001	
17	Plastics	0.00019	–0.00014**	–0.00006*
18	Processed food	–0.00003*	0.00001	0.00000
19	Production technology	–0.00001	–0.00016′	–0.00020*
20	Telecom	–0.00002	–0.00009*	–0.00012**
21	Textiles	0.00010′	–0.00007	–0.00050*
22	Tourism and hospitality	–0.00023**	–0.00082**	–0.00077*
23	Transportation and logistics	–0.00002	0.00001*	0.00012*
	Significantly positive cases (at 5%)	3	3	3
	Significantly negative cases (at 5%)	7	5	12

Note: p-values: ′< 0.1, *< 0.05, **< 0.01 and ***< 0.001.

conditions, for which the correlation also tends to be negative in most industries, with the exceptions of construction and plastics.

Third, the two border-region dummies, which control for a potential underestimation of agglomeration economies across national boundaries, are merely significant. However, growth rates are, not surprisingly, significantly reduced during the years of the financial crisis. Only firms operating in agricultural products and paper products have been shown to be resistant to the macroeconomic recession. Finally, being a subsidiary firm primarily matters for shrinking firms: with the support of a parent company, strong negative growth impulses seem to be cushioned more easily.

Impact of own-industry employment and publications

Having controlled for various firm- and region-specific variables, the impact of the spatially discounted location variables can be discussed. Before taking a closer look at certain peculiarities at the industry level, emerging general patterns are highlighted.

Being located in proximity to employees of the same industry (*AGGL*) does reduce the firms' growth prospect in many industries (Table 3). At the lower quantile $\theta_{0.25}$ this relationship is significantly negative in seven, at the medium quantile $\theta_{0.5}$ in five and at the higher quantile $\theta_{0.75}$ in 12 of the 23 analysed industries. A significant positive relationship is found at each quantile only in three cases.

The prevailing negative agglomeration economies seem to contradict the usual belief that firms benefit from being located in a cluster. This finding may be explained as follows. Firms in clusters benefit from the surroundings in terms of higher innovativeness and competitiveness, but they have to pay higher wages for their employees and higher prices for real estate due to intensive competition (KETELS, 2013). As a consequence, they do not show higher profits and growth rates. Particularly high growth events (at $\theta_{0.75}$) become difficult to realize for incumbent firms in industrial clusters, as many relatively expensive and often highly qualified workers have to be hired. Simpler business activities might even be moved outside of cluster places, where wages are lower, so that the number of employees might even shrink. Besides, firms also become more dependent on the development of surrounding firms that increases vulnerability to industry-specific problems, as the negative effects at $\theta_{0.25}$ indicate.

Exceptions to this pattern exist for construction, distribution, media and publishing and transportation, and logistics, for which agglomeration of own-industry

Table 4. Coefficients for knowledge-generating activities (KNOW) at different quantiles

		KNOW		
ID	Name	$\theta_{0.25}$	$\theta_{0.5}$	$\theta_{0.75}$
1	Agricultural products	–0.00016	–0.00013	–0.00069′
2	Automotive	–0.00031**	–0.00028*	–0.00019
3	Building fixtures and equipment	–0.00026*	–0.00060	–0.00024
4	Business services	0.00039	–0.00069	–0.00055′
			0.00031	
5	Chemical products	–0.00031**	–0.00001	0.00002
6	Construction	–0.00095***	–0.00099***	–0.00120
7	Distribution	0.00015	0.00050	–0.00137
8	Entertainment	0.00064	–0.00001	0.00007
9	Financial services	0.00038	0.00021	0.00061*
10	Heavy machinery	0.00011***	0.00007**	–0.00013*
11	Instruments	0.00045	0.00058	–0.00007
12	Information technology (IT)	0.00061*	0.00045**	0.00037**
13	Media and publishing	0.00022	–0.00008	–0.00076′
14	Medical devices	–0.00048**	0.00019	0.00033**
15	Metal manufacturing	–0.00139	–0.00142**	–0.00140
				–0.00133
16	Paper products	–0.00008	0.00003	0.00006
				0.00027
17	Plastics	0.00018***	–0.00220	–0.00006
		–0.00203		
18	Processed food	0.00077	0.00011	–0.00031
19	Production technology	–0.00064*	–0.00080*	–0.00155*
20	Telecom	–0.00006	0.00018	0.00039*
21	Textiles	0.00017**	–0.00072′	–0.00046
		–0.00127***		
22	Tourism and hospitality	–0.00017	0.00114	0.00196
23	Transportation and logistics	–0.00031*	0.00002	–0.00069
	Significantly positive cases (at 5%)	4	2	4
	Significantly negative cases (at 5%)	8	4	2

Note: p-values: ′< 0.1, *< 0.05, **< 0.01 and ***< 0.001.

employment significantly stimulates the growth performance of firms. This finding can be explained by industry-specific characteristics. In these industries, which are composed of a major part of service activities, value creation strongly depends on frequent interactions with suppliers and customers. Despite negative agglomeration economies in industrial clusters, it seems that benefits from geographic proximity cannot be substituted in these industries by other dimensions of proximity (BOSCHMA, 2005).

Table 4 contains the results for nearby scientific publications (*KNOW*). In eight of the analysed industries, *KNOW* decreases the growth prospects of firms in the lowest quantile ($\theta_{0.25}$). This negative impact, however, diminishing at higher quantiles, with four significant negative cases found at $\theta_{0.5}$ and only two at $\theta_{0.75}$. In contrast hereto, four industries benefit from *KNOW* foremost at $\theta_{0.25}$, whereas different four industries benefit at $\theta_{0.75}$.

These findings might be explained as follows. Growth, stimulated by distance-sensitive knowledge spillovers and learning processes between public research institutes and universities, on the one hand, and private firms, on the other hand, is innovation driven. By providing new market opportunities, product innovations often unleash great potentials for expansion. For process innovations, which imply that the same amount of output can be produced with less input factors, a negative effect on employment growth is often observed in the empirical literature (for an overview, see BUERGER *et al.*, 2012). Hence, innovation-driven growth results in turbulences, as both negative and positive growth events at the tails of the conditional growth rate distribution become more likely.

The positive effect of nearby publications on the growth prospects of firms is found in industries that tend to be more knowledge intensive: financial services, information technology (IT), medical devices, and telecom. In these industries, firms both rely on input from science and provide the necessary absorptive capacity to implement external knowledge. Besides, *KNOW* reduces the likelihood of negative growth events in industries like heavy machinery, plastics or textiles. For these industries, less science-based and innovation-driven in nature, the variable might capture other mechanisms of public research institutes and universities, like graduates, which protect these firms from declining.

Bridging the two measures for economic and knowledge-generating activities by conjointly plotting the estimated coefficients in Fig. 4 shows that the effects

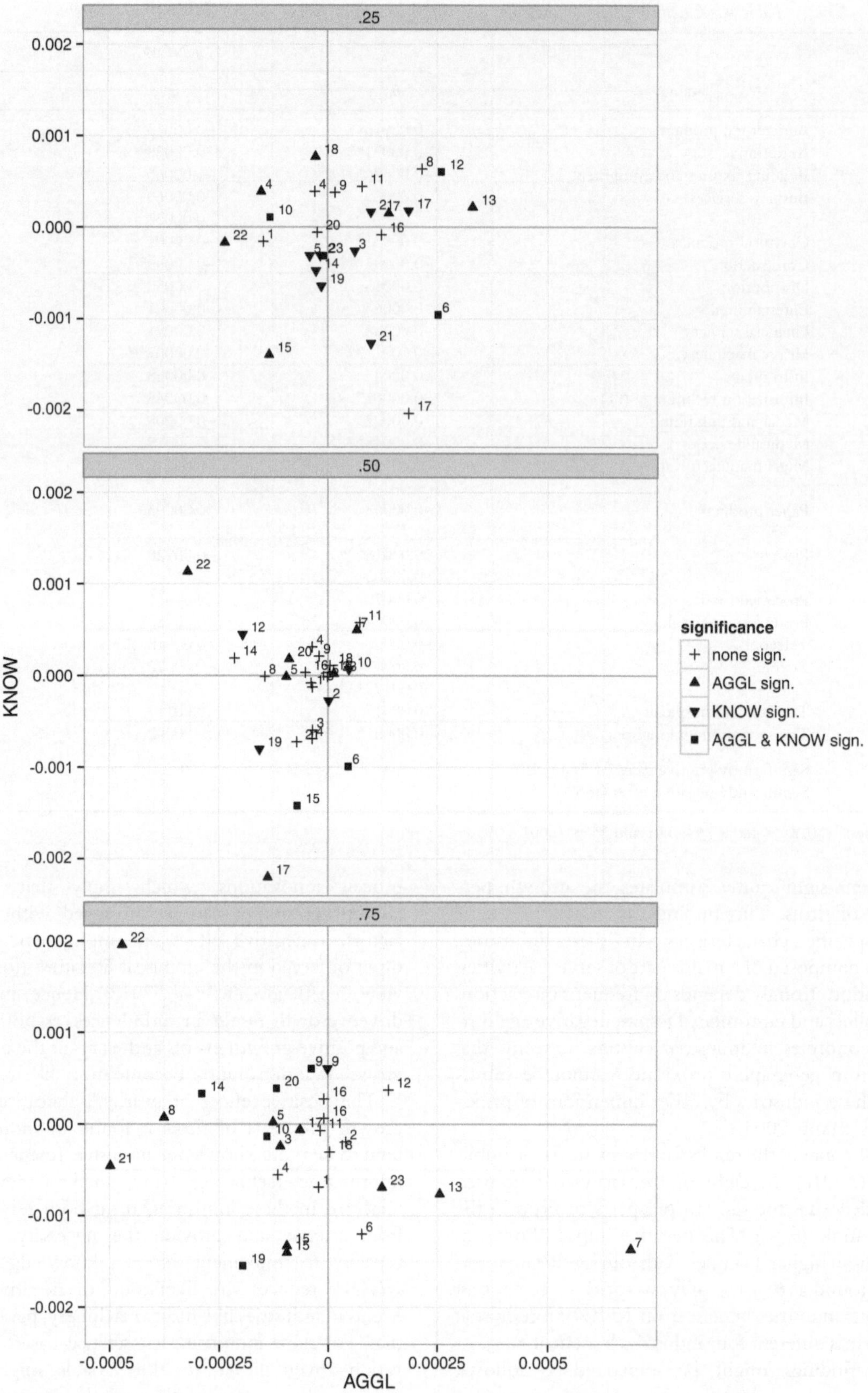

Fig. 4. Scatterplot of coefficients β_1 (AGGL) and β_2 (KNOW) for different quantiles. Significant results (at 5%) are indicated by the symbols

Table 5. Pearson's correlation coefficients of β_1 (AGGL) and β_2 (KNOW) with the industries' age

	$\theta_{0.25}$	$\theta_{0.5}$	$\theta_{0.75}$
β_1 (*AGGL*)	0.158	0.292	0.046
	(0.461)	(0.156)	(0.831)
β_2 (*KNOW*)	−0.400	−0.361	−0.3876
	(0.041)	(0.082)	(0.056)

Note: p-values are given in parentheses.

of *KNOW* and *AGGL* are less pronounced at $\theta_{0.5}$, where the estimates concentrate more closely to the origin, compared with the lower and upper quantiles. This means that the average growing firm is less dependent on economic and knowledge-generating activities in its proximate surrounding, whereas the location matters most for firms in the tails of the conditional growth rate distributions. Besides, a significant negative relationship emerges between the industry-specific estimates of *KNOW* and *AGGL* at $\theta_{0.75}$, with a Pearson's correlation coefficient of−0.411 ($p = 0.046$). The counterbalancing tendencies of the two external factors for strongly expanding firms support the idea that industrial clustering is not an infinite self-reinforcing process.

The second section hypothesized that the actual impact of *AGGL* and *KNOW* on firm growth should depend on the current stage of the industrial life cycle, which can be roughly approximated by the average age of the industries' firms. Correlating this age measure with the corresponding regression coefficients reveals that the older the industry, the less relevant becomes *KNOW* (Table 5). One the one hand, this might be explained by a decreasing importance of inputs from public research institutes and universities in more mature industries (BRENNER and SCHLUMP, 2011), as firms tend to rely more on internal research activities. On the other hand, these firms are increasingly able to invest in and to manage cooperations with the best possible partners, irrespective of the geographic location. Hence, firms of science-based, yet more mature industries, like chemistry, metal manufacturing or production technology, do not benefit from nearby publications. In contrast to expectation, no significant relationship is observed for the industry's age and *AGGL*.

Although some general patterns are identified, a high heterogeneity apparently exists at the industry level: firms of the various industries are not affected by the economic and scientific landscape in the same way. For illustrative reasons, three industries are analysed more in depth: medical devices, distribution and textiles.

The medical devices industry is primarily composed of knowledge-intensive activities. Not yet being a mature industry, locations in proximity to scientific publications increase the firms' likelihood of high growth events. The downside of innovation-driven growth becomes visible at $\theta_{0.25}$, as these firms are simultaneously more prone to decline. The impact of *KNOW* is most pronounced at a spatial scale from 1 to 2 hours, indicating that these effects would have been underestimated by considering regional boundaries that are too narrowly defined. *AGGL* becomes significantly negative at $\theta_{0.75}$ – the upward pressure on wages due to stronger competition in industrial clusters makes it for incumbent firms in the medical devices industry more difficult to grow strongly by hiring many new employees.

Firms belonging to the distribution industry, which is widely composed of wholesale activities, are not affected by *KNOW*. Similar to other service industries, like business services, entertainment, media and publishing, or tourism and hospitality, firms do not rely on scientific input. However, *AGGL* matters all the more: irrespective of the conditional growth rate quantile, firms from distribution strongly benefit from a high agglomeration of own-industry employment. Whilst the impact is most distinctive at short distances of 30 and 21 min for $\theta_{0.25}$ and $\theta_{0.5}$, respectively, highly positive growth events, by contrast, are visible at a larger spatial scale of around 2 hours.

Finally, the textiles industry provides an interesting example in which more than one spatial scale matters at the same time. Looking at the lower quantile of $\theta_{0.25}$, *KNOW* has a positive impact at a distance of 29 min, but a negative one at around 3 hours, implying that different mechanisms might work simultaneously at different spatial scales. Translating this finding into the managers' perspective: optimal location choices require complex spatial multilevel decisions.

CONCLUSIONS

FRENKEN *et al.* (2011) have suggested for future research that one of the main challenges 'lies in settling contradictory empirical findings. In particular […] the main gap in our empirical understanding concerns the effect of localization economies on firm performance, which some may even consider the key question in economic geography at large'. In line with these authors, this paper argues that contradictory empirical findings are closely related to the heterogeneity of firms and industry to which they belong, and of the spatial economic landscape in which they are located.

This paper takes the call for a finer resolution seriously. Several methodological choices are made to account for the omnipresent heterogeneity. First, the approach is micro-geographic in nature, as both the firms and the sources of agglomeration economies are geo-located in space. The unevenly distributed infrastructure that determines the accessibility to these growth relevant external sources is modelled via travel times in the road network, and behavioural assumptions of spatial interactions are reflected by the log-logistic distance-decay function. Moreover, distance-based methods at the micro-geographical level make choices regarding the definition (of the existence) and spatial boundaries of industrial clusters obsolete. As the spatial scale and configuration differ strongly from cluster to cluster (on the spatial clustering of the micro-system technology firms in Germany, see, for instance, SCHOLL and BRENNER, 2012), it is argued that the cluster concept is most meaningfully implemented in cross-sectional, comparative studies by methods that do not rely on a predefined regional (mostly administrative) aggregation level. Second, the cluster concept distinguishes related economic activities from knowledge-generating activities by measuring own-industry employment as well as scientific publications in each municipality. Here, the identification of the best fitting decay function specification is performed for both variables separately, as the 'relevant spatial level and spatial decay may well be different for different mechanisms underlying localization externalities' (FRENKEN *et al.*, 2011, p. 21). Results show that the spatial impact of agglomeration effects are in some cases a sub-regional phenomenon, whilst in other cases they transcend traditionally defined regional boundaries. Third, quantile regression techniques shed light on differences in the relationship between highly growing and declining firms and agglomeration economies. As theorized by HOOGSTRA and VAN DIJK (2004), the spatial surrounding is more influential for firms at the tails of the conditional growth rate distribution. Finally, the disaggregated level of industries accounts for heterogeneity in the underlining technologies and differences along the life cycle stages. Both aspects have increasingly gained attention in the recent literature. The results confirm the existence of differences and particularities when comparing agglomeration economies systematically across industries and support the idea that especially the relevance of external scientific knowledge depends on the industry's age and, hence, on its stage in the life cycle.

Despite the high flexibility of the modelling assumptions, a rather coherent picture of the effects of industrial clusters on the growth performance of incumbent firms emerges. Firstly, being located in agglomerations of related economic activities does not stimulate or even hamper the firms' growth prospects in most industries. This finding is in line with the literature that acknowledges positive cluster effects on start-ups (e.g., SORENSEN and AUDIA, 2000), yet negating the benefits for growth of incumbent firms (FRENKEN *et al.*, 2011). Secondly, proximate knowledge-generating activities, which reflect more directly the specific mechanism of knowledge spillovers, tend to be positively related to firm growth in industries that rely on external scientific knowledge, i.e., knowledge-intensive industries in the early phases of their life cycles. Exceptions of these general patterns exist, and often can be explained by taking a closer look at properties and particularities of the respective industries.

The results have important implications for both policy-makers and firm managers. The growth dynamics of incumbent firms are generally seen as a key driver for the creation and elimination of job opportunities in regional labour markets. In most industries, policy measures that foster further clustering of economic activities are confronted with prevailing diseconomies of agglomeration. Depending on the kind of industrial specialization, scarce resources might be more effectively invested in activities of universities or public research institutes. However, due to increased competition, wages and, thus, higher value added and taxes, it is the region and the people that benefit most from the agglomeration of economic activities, making industrial clusters a relevant policy issue. But also firm managers might improve their growth prospects and the related market expansion objectives by taking the economic environment (at different spatial scales) into account when deciding on the location of headquarters.

Like BEAUDRY and SWANN (2009), this paper focuses on the differences of agglomeration economies among industries. Future research could explore more systematically the reasons behind the industry-specific differences, for instance by employing measures on the knowledge properties of the industries or by disentangling the effects and mechanisms in comparative case studies. Complementary hereto, an emerging body of literature investigates the differences that can be observed among firm-specific categories, such as their size or age (e.g., DUSCHL *et al.*, 2014; NEFFKE *et al.*, 2012; RIGBY and BROWN, 2013). Hence, a stronger integration of these related streams of literature is needed.

Acknowledgements – The authors would like to thank Martin Andersson and Ron Boschma as well as the participants of the 'Geography of Innovation' Conference 2012 in St Etienne, and of the International PhD School on Economic Geography 2012 in Utrecht for helpful comments and suggestions on an earlier draft of this paper. This paper has also benefited from detailed comments of two anonymous referees. The usual disclaimers apply.

NOTES

1. This size restriction implies that 2.2% of total firms in the database are dropped from the sample. Also, a different threshold of 500 employees (with a loss of 4.5%) was tested; however, the results remain robust.
2. By repeating the analysis on a year-by-year basis (for industries with enough observations) it can be shown that the results remain robust.
3. A quadratic term to control for non-linear relationships was initially included, but never found to be significant.
4. Performing an extensive Monte Carlo analysis, these authors show that this approach captures substantive spatial dependence in the dependent variable and accounts for both local and global spillovers.
5. The detailed regression results of the control variables are available from the authors upon request.

REFERENCES

ANDERSSON M. and GRASJÖ U. (2009) Spatial dependence and the representation of space in empirical models, *Annals of Regional Science* **43**, 159–180.

ANDERSSON M. and KARLSSON C. (2007) Knowledge in regional economic growth – the role of knowledge accessibility, *Industry and Innovation* **14**, 129–149.

ANDERSSON M. and LÖÖF H. (2011) Agglomeration and productivity: evidence from firm-level data, *Annals of Regional Sciences* **46**, 601–620.

ARTIS M. J., MIGUELEZ E. and MORENO R. (2012) Agglomeration economies and regional intangible assets: an empirical investigation, *Journal of Economic Geography* **12**, 1167–1189.

AUDRETSCH D. B. and DOHSE D. (2007) Location: a neglected determinant of firm growth, *Review of World Economics* **143(1)**, 79–107.

AUDRETSCH D. B. and FELDMAN M. P. (1996) Innovative clusters and the industry life cycle, *Review of Industrial Organization* **11**, 253–273.

AUDRETSCH D. B. and LEHMANN E. E. (2005) Mansfield's missing link: the impact of knowledge spillovers on firm growth, *Journal of Technology Transfer* **30(1)**, 207–210.

BALDWIN J., BECKSTEAD D., BROWN W. M. and RIGBY D. (2008) Agglomeration and the geography of localization economies in Canada, *Regional Studies* **42**, 117–132.

BALDWIN J., BROWN W. M. and RIGBY D. (2010) Agglomeration economies: microdata panel estimates from Canadian manufacturing, *Journal of Regional Science* **50(5)**, 915–934.

BEAUDRY C. and SWANN P. G. M. (2009) Firm growth in industrial clusters of the United Kingdom, *Small Business Economics* **32**, 409–424.

BEUGELSDIJK S. (2007) The regional environment and a firm's innovative performance: a plea for a multilevel interactionist approach, *Economic Geography* **83(2)**, 181–199.

BOSCHMA R. (2005) Proximity and innovation: a critical assessment, *Regional Studies* **39(1)**, 61–74.

BOSCHMA R., ERIKSSON R. and LINDGREN U. (2009) How does labour mobility affect the performance of plants? The importance of relatedness and geographical proximity, *Journal of Economic Geography* **9**, 169–190.

BRENNER T. and SCHLUMP C. (2011) Policy measures and their effects in the different phases of the cluster life cycle, *Regional Studies* **45(10)**, 1363–1386.

BUCHINSKY M. (1998) Recent advances in quantile regression models, *Journal of Human Resources* **33**, 88–126.

BUENSTORF G. and KLEPPER, S. (2009) Heritage and agglomeration: the Akron tyre cluster revisited, *Economic Journal* **119(537)**, 705–733.

BUERGER M., BROEKEL T. and COAD A. (2012) Regional dynamics of innovation: investigating the co-evolution of patents, research and development (R&D), and employment, *Regional Studies* **46**, 565–582.

BUERGER M., VAN OORT F. G. and VAN DER KNAAP B. (2010) A treatise on the geographical scale of agglomeration externalities and the MAUP, *Scienze Regionali* **9(1)**, 19–40.

CAINELLI G. (2008) Spatial agglomeration, technological innovations, and firm productivity: evidence from Italian industrial districts, *Growth and Change* **39(3)**, 414–435.

CASSIA L., COLOMBELLI A. and PALEARI S. (2009) Firms' growth: does innovation system matter?, *Structural Change and Economic Dynamics* **20**, 211–220.

COAD A. (2009) *The Growth of Firms: A Survey of Theories and Empirical Evidence.* Edward Elgar, Cheltenham.

COAD A. (Forthcoming 2014) The meso-level: structural persistency, in CANTNER U. and GUERZONI M. (Eds) *Handbook on Industrial Dynamics and Evolutionary Economics.* Edward Elgar, Cheltenham.

COHEN S., DI MININ A., MOTOYAMA Y. and PALMBERG C. (2009) The persistence of home bias for important R&D in wireless telecom and automobiles, *Review of Policy Research* **26(1–2)**, 55–79.

DE VRIES J., NIJKAMP P. and RIETVELD P. (2009) Exponential or power distance – decay for commuting? An alternative specification, *Environment and Planning A* **41(2)**, 461–480.

DELGADO M., PORTER M. E. and STERN S. (2010) Clusters and entrepreneurship, *Journal of Economic Geography* **10**, 495–518.

DELGADO M., PORTER M. E. and STERN S. (2012) *Clusters, Convergence, and Economic Performance.* NBER Working Paper Number 18250. National Bureau of Economic Research (NBER), Cambridge, MA.

DOSI G., LECHEVALIER S. and SEECHI A. (2010) Introduction: Interfirm heterogeneity – nature, sources and consequences for industrial dynamics, *Industrial and Corporate Change* **19(6)**, 1867–1890.

DRUCKER J. (2012) *The Spatial Extent of Agglomeration Economies: Evidence from Three U.S. Manufacturing Industries.* Center for Economic Studies Number 12–01.

DRUCKER J. and FESER E. (2012) Regional industrial structure and agglomeration economies: an analysis of productivity in three manufacturing industries, *Regional Science and Urban Economics* **42**, 1–14.

DUNNING J. and LUNDAN S. (2009) The internationalization of corporate R&D: a review of the evidence and some policy implications for home countries, *Review of Policy Research* **26(1–2)**, 13–33.

DURANTON G. and OVERMAN H. G. (2005) Testing for localization using micro-geographic data, *Review of Economic Studies* **72**, 1077–1106.

DUSCHL M., SCHIMKE A., BRENNER T. and LUXEN D. (Forthcoming 2014) Firm growth and the spatial impact of geolocated external factors – empirical evidence for German manufacturing firms, *Jahrbucher für Nationalökonomie und Statistik.*

Eckey H., Kosfeld R. and Türck M. (2006) Abgrenzung deutscher Arbeitsmarktregionen, *Raumforschung und Raumordnung* **64**, 299–309.

Eriksson R. H. (2011) Localized spillovers and knowledge flows: how does proximity influence the performance of plants? *Economic Geography* **87(2)**, 127–152.

Evans D. H. (1987) The relationship between firm growth, size, and age: estimates for 100 manufacturing industries, *Journal of Industrial Economics* **35(4)**, 567–581.

Feldman M. P. (1999) The new economics of innovation, spillovers and agglomeration: a review of empirical studies, *Economics of Innovation and New Technology* **8(1–2)**, 5–25.

Fornahl D., Broekel T. and Boschma R. (2011) What drives patent performance of German biotech firms? The impact of R&D subsidies, knowledge networks and their location, *Papers in Regional Science* **90(2)**, 395–418.

Frenken K., Cefis E. and Stam E. (2011) *Industrial Dynamics and Economic Geography: A Survey*. Working Paper Number 11.07. Eindhoven Centre for Innovation Studies (ECIS), Eindhoven.

Frenken K., van Oort F. and Verburg T. (2007) Related variety, unrelated variety and regional economic growth, *Regional Studies* **41(5)**, 685–697.

Geisberger R., Luxen D., Neubauer S., Sanders P. and Volker L. (2010) Fast detour computation for ride sharing, in Erlenbach T. and Lübbecke M. (Eds) *Proceedings* of the 10th Workshop on Algorithmic Approaches for Transportation Modelling, Optimization, and Systems (ATMOS'10), pp. 88–99.

Glaeser E. L., Kallal H. D., Scheinkman J. A. and Schleifer A. (1992) Growth in cities, *Journal of Political Economy* **100(6)**, 1126–1152.

Gould W. (1992) Quantile regression with bootstrapped standard errors, *Stata Technical Bulletin* **9**, 19–21.

Graham D. J. (2009) Identifying urbanisation and localisation externalities in manufacturing and service industries, *Papers in Regional Science* **88(1)**, 63–94.

Graham D. J., Gibbson S. and Martin R. (2010) *The Spatial Decay of Agglomeration Economies: Estimates for Use in Transport Appraisal.* Final Report. LSE Imperial College, London.

Guimón J. (2009) Government strategies to attract R&D-intensive FDI, *Journal of Technological Transfer* **34(4)**, 364–379.

Guiso L. and Schivardi F. (2007) Spillovers in industrial districts, *Economic Journal* **117**, 68–93.

Harris R. (2011) Models of regional growth: past, present and future, *Journal of Economic Survey* **25(5)**, 913–951.

Henderson J. V. (2003) Marshall's scale economies, *Journal of Urban Economies* **53**, 1–28.

Henderson J. V., Kuncoro A. and Turner M. (1995) Industrial development in cities, *Journal of Political Economy* **103(5)**, 1067–1090.

Hoogstra G. J. and van Dijk J. (2004) Explaining firm employment growth: does location matter?, *Small Business Economics* **22**, 179–192.

Hymer S. and Pashigian P. (1962) Firm size and rate of growth, *Journal of Political Economy* **70(6)**, 556–569.

Jaffe A. M. (1989) Real effects of academic research, *American Economic Review* **79**, 957–970.

Johansson B., Klaesson J. and Olsson M. (2003) Commuters' non-linear response to time distances, *Journal of Geographical Systems* **5**, 315–329.

Ketels C. (2013) Recent research on competitiveness and clusters: what are the implications for regional policy?, *Cambridge Journal of Regions, Economy and Society* **6(2)**, 269–284.

Knopp S., Sanders P., Schultes D., Schulz F. and Wagner D. (2007) Computing many-to-many shortest paths using highway hierarchies. Paper presented at the Workshop on Algorithm Engineering and Experiments, January 2007.

Koenker R. (2005) *Quantile Regression.* Cambridge University Press, New York, NY.

Koo J. (2005) Agglomeration and spillovers in a simultaneous framework, *Annals of Regional Science* **39**, 35–47.

Kubis A., Brachert M. and Titze M. (2009) *Industrielle Cluster als Ursache regionaler Prosperität? Zur Konvergenz deutscher Arbeitsmarktregionen 1996–2005.* Institut für Wirtschaftsforschung Halle (IWH), Halle.

Lychagin S., Pinske J., Slade M. and van Reenen J. (2010) *Spillovers in Space: Does Geography Matter?* NBER Working Paper Series Number 16188. National Bureau of Economic Research (NBER), Cambridge, MA.

Malmberg A., Malmberg B. and Lundquist P. (2000) Agglomeration and firm performance: economies of scale, localisation, and urbanisation among Swedish export firms, *Environment and Planning A* **32**, 305–321.

Neffke F., Henning M. and Boschma R. (2012) The impact of aging and technological relatedness on agglomeration externalities: a survival analysis, *Journal of Economic Geography* **12**, 485–517.

Neffke F., Henning M., Boschma R., Lundquist K.-J. and Olander, L.-O. (2011) The dynamics of agglomeration externalities along the life cycle of industries, *Regional Studies* **45(1)**, 49–65.

Openshaw S. (1984) *The Modifiable Areal Unit Problem.* Geo Books, Norwich.

Otto A. and Fornahl D. (2008) *Long-Term Growth Determinants of Young Businesses in Germany: Effects of Regional Concentration and Specialisation.* IAB Discussion Paper Number 13/2008. Institut für Arbeitsmarkt- und Berufsforschung (IAB), Nuremberg.

Pinske J. and Slade M. (2010) The future of spatial econometrics, *Journal of Regional Science* **50**, 103–117.

Porter M. E. (2000) Location, competition, and economic development: local clusters in a global economy, *Economic Development Quarterly* **14**, 15–34.

Porter M. E. (2003) The economic performance of regions, *Regional Studies* **37(6–7)**, 549–578.

Potter A. and Watts H. D. (2011) Evolutionary agglomeration theory: increasing returns, diminishing returns, and the industry life cycle, *Journal of Economic Geography* **11**, 417–455.

Raspe O. and van Oort F. (2011) Growth of new firms and spatially bounded knowledge externalities, *Annals of Regional Science* **46**, 495–518.

RIGBY D. L. and BROWN J. (2013) Who benefits from agglomeration?, *Regional Studies* DOI: 10.1080/00343404.2012.753141.

RIGBY D. L. and ESSLETZBICHLER J. (2002) Agglomeration economies and productivity differences in US cities, *Journal of Economic Geography* **2**, 407–432.

RODRÍGUEZ-POSE A. and CRESCENZI R. (2008) Research and development, spillovers, innovation systems, and the genesis of regional growth in Europe, *Regional Studies* **42(1)**, 51–67.

ROSENTHAL S. S. and STRANGE W. C. (2003) Geography, industrial organization, and agglomeration, *Review of Economics and Statistics*, **85(2)**, 377–393.

SCHOLL T. and BRENNER T. (2012) *Detecting Spatial Clustering Using a Firm-Level Cluster Index*. Marburg Working Papers on Innovation and Space Number 02.12. Marburg.

SORENSEN O. and AUDIA P. G. (2000) The social structure of entrepreneurial activity: geographic concentration of footwear production in the United States, 1940–1989, *American Journal of Sociology* **106(2)**, 424–262.

SPENCER G. M., VINODRAI T., GERTLER M. S. and WOLFE D. A. (2010) Do clusters make a difference? Defining and assessing their economic performance, *Regional Studies* **44(6)**, 697–715.

THORSEN I., UBOE J. and NAVDAL G. (1999) A network approach to commuting, *Journal of Regional Science* **39(1)**, 73–101.

UNITED NATIONS CONFERENCE ON TRADE AND DEVELOPMENT (UNCTAD) (2005) *World Investment Report 2005*. UNCTAD, Geneva.

VAN SOEST D. P., GERKING S. and VAN OORT F. G. (2006) Spatial impacts of agglomeration externalities, *Journal of Regional Science* **46(5)**, 881–899.

WENNBERG K. and LINDQVIST G. (2010) The effect of clusters on the survival and performance of new firms, *Small Business Economics* **34**, 221–241.

Marshall's versus Jacobs' Externalities in Firm Innovation Performance: The Case of French Industry

DANIELLE GALLIANO, MARIE-BENOÎT MAGRINI and PIERRE TRIBOULET
INRA, Castanet Tolosan, France

GALLIANO D., MAGRINI M.-B. and TRIBOULET P. Marshall's versus Jacobs' externalities in firm innovation performance: the case of French industry, *Regional Studies*. This paper analyses the influence of different types of spatial externalities related to the location of firms on their innovation performance and how those externalities combine in the territories with regard to the Marshall–Jacobs dichotomy. The originality of this study also lies in the consideration of a larger definition of the firm, one that takes into account the location of all its units. Based on a dataset of French industrial firms and specific indicators to evaluate the specialization and/or diversification of the employment zones, the impacts of the spatial profile of the firm on its innovation performance are tested.

GALLIANO D., MAGRINI M.-B. and TRIBOULET P. 企业创新表现中的马歇尔外部性对雅各布斯外部性：法国工业的案例，*区域研究*。本文分析与企业区位有关的不同类型的空间外部性，对企业的创新表现产生的影响，以及这些外部性在马歇尔—雅各布斯的二元对立中，如何在领域中相互结合。本文的创新性，亦在于将企业的所有单位之区位皆纳入考量，以更广泛地考量企业的定义。本研究根据一个法国工业企业的数据集和评估就业区的专殊化以及／或多样化特定指标，测试企业的空间形态对创新表现的影响。

GALLIANO D., MAGRINI M.-B. et TRIBOULET P. Les externalités Marshall versus Jacobs dans la performance innovatrice de la firme: le cas de l'industrie française, *Regional Studies*. Ce papier analyse l'influence des externalités spatiales en lien avec la localisation de la firme sur sa performance à l'innovation et comment ces externalités spatiales se combinent dans les territoires au regard de la dichotomie Marshall–Jacobs. L'originalité de notre étude réside également dans la prise en compte d'une définition élargie de la firme qui intègre la localisation de tous ses établissements. Basé sur un ensemble de données d'entreprises industrielles françaises et deux indicateurs pour évaluer la spécialisation et/ou la diversification des zones d'emploi en France, nous testons l'impact du profil spatial de la firme sur sa performance innovatrice.

GALLIANO D., MAGRINI M.-B. und TRIBOULET P. Marshall- und Jacobs-Externalitäten bei der Innovationsleistung von Firmen: der Fall der französischen Industrie, *Regional Studies*. In diesem Beitrag wird der Einfluss verschiedener Arten von räumlichen Externalitäten im Zusammenhang mit dem Standort von Firmen auf ihre Innovationsleistung analysiert, und es wird untersucht, wie sich diese Externalitäten in den Gebieten hinsichtlich der Marshall-Jacobs-Dichotomie miteinander kombinieren. Die Originalität dieser Studie liegt auch in der Verwendung einer breiteren Definition des Begriffs der Firma begründet, bei der der Standort von allen ihren Niederlassungen berücksichtigt wird. Ausgehend von einem Datensatz französischer Industriefirmen und von spezifischen Indikatoren zur Bewertung der Spezialisierung und/oder Diversifizierung der Beschäftigungszonen werden die Auswirkungen des räumlichen Profils der Firma auf ihre Innovationsleistung überprüft.

GALLIANO D., MAGRINI M.-B. y TRIBOULET P. Externalidades de Marshall y Jacobs en el desempeño de innovación de las empresas: el caso de la industria francesa, *Regional Studies*. En este artículo analizamos la influencia de los diferentes tipos de externalidades espaciales con relación a la ubicación de las empresas en cuanto a su desempeño de innovación y cómo se combinan estas externalidades en los territorios con respecto a la dicotomía Marshall–Jacobs. La originalidad de nuestro estudio radica en considerar una definición más amplia del concepto de empresa que tiene en cuenta la ubicación de todas sus filiales. A partir de un grupo de datos de empresas industriales francesas y de indicadores específicos para evaluar la especialización y/o diversificación de las zonas de empleo, comprobamos los efectos del perfil espacial de la empresa con respecto a su desempeño de innovación.

INTRODUCTION

The economic literature on the spatial determinants of innovation has largely emphasized the fact that the environment in which a firm is located influences its ability to innovate or to adopt innovations (BOSCHMA, 2005; MASSARD and MEHIER, 2009; FELDMAN and KOGLER, 2010, for a survey). Also, a classical question in economic geography is to evaluate how spatial agglomeration contributes to the economic performance of firms and to their innovativeness, with the idea being that agglomeration economies enhance the cross-fertilization of ideas and technologies, i.e. foster knowledge externalities (ROSENTHAL and STRANGE, 2003; AUDRETSCH and FELDMAN, 2004; AUTANT-BERNARD and LESAGE, 2011). As recalled by NEFFKE *et al.* (2011), the literature on agglomeration economies has, since the publication of JACOBS' (1969) works and MARSHALL's (1890) studies, distinguished between the two types of externalities according to the idea that diversified cities offer different benefits compared with specialized cities. However, as many authors note, the literature remains inconclusive in determining which of the two concepts provides a more favourable environment for innovation (VAN DER PANNE and BEERS, 2006; BEAUDRY and SCHIFFAUEROVA, 2009; NEFFKE *et al.*, 2011). On the other hand, it is generally recognized that knowledge spillovers are geographically bounded within a limited space and that the attempts to understand the micro-foundations of agglomeration economies need to be further developed (ROSENTHAL and STRANGE, 2003; PUGA, 2010).

Specifically, the emphasis on the spatial effects on innovation often rests on a limited conception of firm spatial organization, and most studies take into consideration only the location of the firm head offices. However, certain studies have highlighted that the other units of a firm could play an important role in its performance and, for instance, in its propensity to adopt new technologies (BEARDSELL and HENDERSON, 1999; GALLIANO *et al.*, 2011). Thus, while the influence of head office location on the propensity of the firm to innovate has often been highlighted, taking into account the environment where the other units of the firm are located could help to evaluate better its innovation capacity in relation to the key role played by knowledge externalities. Indeed, according to whether the spatial externalities are captured through the head office alone or also through the other units, the impact of the spatial environment of the firm on its innovation capacity could change.

The main objective of this paper is to test the influence of the spatial externalities on firm innovation performance and, more precisely, to determine which type of the spatial externalities is the most effective with regard to the Marshall–Jacobs dichotomy. The first innovative contribution is to test the potential joint effect of both types of the externalities that the literature tends to separate. The second contribution is to take into account a larger definition of the firm (not limited to the sole location of the head office) to improve the evaluation of the impact of spatial externalities on innovation.

To make empirical progress on this issue, the 2008 Community Innovation Survey is used, which provides in-depth information regarding product innovation in terms of the decision-making and intensity for almost 5000 firms and their 17000 units, representative of all French manufacturing firms. The spatial externalities are captured at the level of the employment areas (304 areas in France) using the location quotient (LQ) and Herfindahl indices, which are largely used in the agglomeration literature to characterize industrial specialization and diversification, respectively. These indices are calculated using an exhaustive database on the employment in the manufacturing industry units. One can subsequently test the causal relationships between the specialized and diversified externalities and the degree of firm innovativeness by considering the spatial environment of the firm at the head office. To proceed further with this issue, it is proposed to characterize the spatial environment of the firm as the one that concerns the majority of the firm's employees and not only the head office. In this way, one can address the question of whether the positive effect of agglomeration economies, largely demonstrated at the head office level, can also impact the other components of the firm.

To conduct this analysis, three main models have been constructed and estimated, and the results indicate that the geographic environment of the firm does not have identical effects on the probability to innovate and on the innovation intensity and that these effects also vary depending on the spatial organizational structure. The first model, called the 'diversification model', tends to reveal the stimulating effect of the 'Jacobian' diversity on a firm's decision to engage in innovation and the negative effect on the innovation intensity. The second model, called the 'specialization model', demonstrates that being located in the specialized zones reinforces innovation intensity. Finally, the third model, denoted as the 'dual model', emphasizes the combined effects of specialization and diversification on innovation.

The article is structured as follows. The following section discusses the influence of the agglomeration economies on firm innovation performance relative to the debate on the specialization versus diversification externalities. The paper then presents the data and methodology, the spatial environment variables and the econometric models used to test the impact of these types of externalities on product innovation according to the two perspectives of the firms. The fourth section presents the results; and the last section summarizes the primary conclusions.

INNOVATION AND SPATIAL EXTERNALITIES: THE DEBATE CONCERNING THE SPECIALIZATION VERSUS DIVERSIFICATION EFFECTS

Agglomeration economies and innovation performance

In the analysis of agglomeration economies, the externalities play a central role. The externalities related to the geographical environment for the most part refer to the external economies generated by the geographical concentration of firms. As noted in the introduction, two types of agglomeration economies are traditionally highlighted in the literature (DICKEN and MALMBERG, 2001). Agglomeration economies are the result either of a geographical concentration of similar industrial activities linked to the economies of a location or of a diversified concentration of activities more strongly linked to the urban economies.

Economies of location are typically related to the degree of industrial specialization in the area where the firm is located. The local specialization enables firms to develop a network of dedicated suppliers and to gain access to a specialized workforce; they facilitate the diffusion of information and innovation between competing firms, and the economies of scale realized in one sector are related to the presence of a larger labour market and of a collective infrastructure. This concept, developed by MARSHALL (1890), ARROW (1962), ROMER (1986) and later formalized by GLAESER *et al.* (1992), became known as the Marshall–Arrow–Romer (MAR) agglomeration economies model. As summarized by BEAUDRY and SCHIFFAUEROVA (2009), 'this model claims that the concentration of an industry in a region promotes knowledge spillovers between firms and facilitates innovation in that particular industry within the region' (p. 318). This model rests on the idea that knowledge externalities only occur between firms of the same industry and therefore that they can only be facilitated by the geographical concentration of firms that belong to similar industries. This is the reason that the intra-industry spillovers are usually called localization or specialization or MAR externalities. These terms will be used interchangeably.

On the other hand, the urban economies are 'local diversity related' and facilitate the access to a qualified and diversified workforce and to a variety of infrastructures, community facilities and activities. They depend fundamentally on the size of the cities. Indeed, JACOBS (1969) argues that the most important sources of knowledge spillovers are external to the industry within which the firm operates. Unlike Marshall, Jacobs emphasizes that it is above all the local variety of industries that promotes knowledge spillovers and ultimately innovation activity. As BEAUDRY and SCHIFFAUEROVA (2009) recall, 'a more diverse industrial fabric in close proximity fosters opportunity to imitate, share and recombine ideas and practices across industries' (p. 319). The exchange of complementary knowledge fosters research and the experimentation for innovation, and a more diversified economy will contribute to these complementarity knowledge bases. These inter-industry spillovers are known as urbanization,[1] diversification, or Jacobs externalities. As for the MAR externalities, these three terms will be used indistinctively.

Though Marshall and Jacobs agree that the concentration of firms generates geographic effects that impact their innovation capacities, they disagree on the effect of industry concentration: According to the MAR concept, knowledge spillovers occur between firms that operate in similar industries; Jacobs argues otherwise. And it is clear, as VAN DER PANNE (2004) noted, that there is still debate as to whether agglomeration economies occur between firms that belong to the same sector rather than between firms that operate in different industries. As Marshall emphasized, knowledge is predominantly industry-specific and knowledge spillovers arise between firms in the same industry. Others argue that diversification rather than specialization externalities foster firm innovative performance and their agglomeration (AUDRESTCH and FELDMAN, 2004). Thus, there is no unanimity in the literature about the impact of the agglomeration effects (FU and HONG, 2011), and it seems, as argued by FRENKEN *et al.* (2007), that within one type of geographic area, both the effects related to specialization and those related to diversity can be observed.

However, as finally explained in BEAUDRY and SCHIFFAUEROVA's (2009) review of the literature, it seems that this debate is strongly linked to the methodology used to measure the levels of specialization and diversification of areas. Based on 67 reviewed articles, these authors explain that 'the 3-digit industrial and geographical aggregation together with the choice of performance measures, specialisation and diversity indicators are the main causes for the lack of resolution in the debate' (p. 318). Thus, an examination of the indices used to measure these two types of externalities is proposed.

Specialization versus diversification measures of agglomeration economies

As explained above, this paper will consider the fact that a zone is specialized as the basis for identifying the MAR externalities and the fact that a zone is diversified as evidence of the existence of Jacobs externalities. The presence of both types of externalities is not tested with the same indicator. In the literature, the most commonly used indicator of the MAR externalities is the LQ, which represents the relationship between an area's share of a particular industry and the national share. This share is, in most cases, measured on the basis of employment.

There exists a large variety of approaches to measuring the Jacobs externalities, but the most commonly used indicator remains the Herfindahl index. It is calculated as the weight of a given sector in a given geographic area, taking into account the total weight of the sector in the entire territory. The second most employed indicator is the Gini index, which is calculated in a similar way to the Herfindahl index. Note that the literature tends to distinguish between the indicators that measure *diversity* per se and those that evaluate the *market size*, with the idea that diversity increases with the size of the agglomeration area. The Herfindahl index is considered as a more diversity-based measure, whereas the total employment rate (in the industrial sector or in both the industrial and service sectors) is used as a proxy of the size of the agglomerations. The latter is more appropriate to capture the global urbanization externalities than the specific industrial diversity of an area.

The value of these indicators is strongly related to the level of aggregation chosen both as regards the industrial classification and the geographical units. With respect to the industry classification, the census of BEAUDRY and SCHIFFAUEROVA (2009) indicates that the presence of Jacobs externalities is more easily detected with a more detailed industry classification, whereas the probability of detecting the MAR externalities is higher with a more aggregated classification. The literature also presents the various effects of specialization or diversification economies according to the nature of the sector – for instance, the manufacturing versus service sector. Cross-fertilization and spillovers may therefore influence sectors differently and thus prompt the authors to conduct analyses based on activities – such as the manufacturing sector – that are strongly represented in the territories. The level of geographical aggregation could also influence the probabilities of detecting the MAR or Jacobs externalities. It is recognized that dividing a territory on the basis of its economic characteristics is more appropriate than a strictly administrative division. The labour zones comprising the municipalities and determined on the basis of home-to-work commuting flows serve as the relevant economic units for measuring the knowledge externalities across industries. This division is also interesting as it is applied to the entire territory of a country. This contrasts with other studies, which are based on large cities and fail to take into account a large part of the economic activity. Indeed, BEAUDRY and SCHIFFAUEROVA's (2009) census reveals that 'with smaller geographical units, there are more of Marshall and Jacobs simultaneous positive results' (p. 326). This principle was also observed by GLAESER *et al.* (1992). However, as explained above, this makes it difficult to distinguish what is related to a sector-based division and to a geographic division.

The definition of the firm

Other aspects are likely to contribute to the controversy regarding specialization and diversification, particularly those related to the firm characteristics. Indeed, various authors have highlighted that the firm characteristics influence the benefits they draw from the agglomeration effects. As FELDMANN and KOGLER (2010) noted, a fundamental question concerns what types of firms are able to absorb the knowledge spillovers and to benefit from them. MCCANN and FOLTA (2011) demonstrate, among other things, that the younger firms and the firms with higher knowledge stocks benefit more from geographic agglomeration. They also make the distinction between the agglomeration of industrial goods manufacturers that benefit from supply related externalities and that of the retailers of consumer goods that benefit from demand-related externalities. Furthermore, NEFFKE *et al.* (2011) demonstrate that through the evolution from the early to the intermediate and then to the mature stages, the industries benefit more from the derived local specialization that steadily increases and that, on the contrary, the benefits from local diversity are positive for the young industries. Finally, these recent studies on economic geography underline the importance of the firm characteristics in understanding the benefit to be drawn from their environment (FELDMAN and KOGLER, 2010).

The spatial organization of the firm. In this line of thought, one purpose of this article is to take into account the complexity of the firm spatial organization, a complexity that can be related to the location of several of its units in different types of areas. Having several units enables a functional separation of activities and a spatial division of labour, which give the firm the opportunity to efficiently manage a number of locations, and thus the firm could benefit from the various types of externalities (OTA and FUJITA, 1993; WOOD and ROBERTS, 2011). Thus, in the Porterian 'home base' traditional framework (PORTER, 1990), the head office is usually defined as the locus of overall control over the firm and as the centre of strategic decision making, while the multi-location structure allows a functional separation of activities. Moreover, a locational hierarchy between the types of places could exist (WOOD and ROBERTS, 2011). A firm can then perform an activity using the different spatial environments in which it operates, that is, the environments that are each likely to have a unique influence on the firm's capacity to innovate (AUDRESTCH and FELDMAN, 2004; MAGRINI and GALLIANO, 2012). Moreover, as AUDIA *et al.* (2001) have demonstrated, the multi-unit location of a firm, per se, can foster learning between its units and facilitate organizational innovation. GALLIANO *et al.* (2011) also demonstrate that this multi-location structure is conducive to technology adoption. Thus, if the spatial externalities can influence a firm's innovation capacity and particularly if it is recognized that the high level of innovation of a firm is generally associated with the location of the head offices in urban areas, the question becomes what impact the multiple locations of a firm have on its propensity to innovate.

Other main determinants of a firm's innovation. Moreover, the literature has identified several other determinants of a firms' propensity to innovate related to its structural characteristics, its relationship with external actors and its commercial and competitive environments. With regard to the structural characteristics, the overall idea is that the firm must be able to rely on its internal knowledge bases and to access and absorb external knowledge to innovate (COHEN and LEVINTHAL 1989). Thus, firms have their own specific characteristics, such as size, R&D expenditures, the decision-making and knowledge-diffusion organization, which influence the propensity to innovate (the 'rank effects' mentioned by KARSHENAS and STONEMAN, 1993). Thus, a large size, belonging to a group and a high R&D intensity have a positive influence on the firm's innovation capacity and, above all, can increase its absorption capacity and help the firm overcome the difficulties experienced during the innovation process (LHUILLERY and PFISTER, 2009). Furthermore, innovation does not occur independently from the firm's internal resources, in keeping with the idea of complementarity with other types of organizational innovation and change developed by MILGROM and ROBERTS (1990). Thus, the lack of associated specific resources hinders not only the success of the product innovation but also the decision to engage in an innovation process (RAMA and VON TUNZELMANN, 2008). Finally, the sector, through its technological intensity and appropriability conditions, could also have a strong influence (VON TUNZELMANN and ACHA, 2005).

Concerning the external relationships, the external sources of information selected by the firm (incoming spillovers), as well as its choices in terms of the cooperation with external partners, could have an important influence on firm innovation. For instance, ANDERSEN and LUNDVALL (1988) demonstrate that a firm's product and process innovation depends not only on its R&D investments but also on the degree of interaction with its suppliers of equipment and specialized inputs. Moreover, the cooperation patterns with competitors are more complex than with the upstream and downstream partners, particularly because they are more difficult to manage. Other studies find that cooperation with universities and research organizations has a non-significant or even negative effect on innovation (LHUILLERY and PFISTER, 2009). Finally, other variables characterizing the competitive environment – particularly the degree of market concentration or the openness to international markets – can have a structuring and a positive effect on a firm's propensity to innovate. In our empirical model, the latter determinants will be considered as variables controlling for the effect of the firm's spatial profile.

Finally, the emphasis on the spatial effects on innovation often rests on a limited conception of firm spatial organization, and most studies only take into consideration the location of the firm head offices. Nevertheless, certain studies have highlighted that the spatial organization of a firm plays an important role in innovation (MAGRINI and GALLIANO, 2012). Thus, while the influence of the head office's location on the firm's propensity to innovate has often been highlighted, taking into account the other units of the firm and a larger spatial environment could help to better understand the role played by spatial externalities in terms of whether the latter are captured through the head office only or also through the other units. The proposal is then to test the influence of spatial externalities on firm innovation according to two conceptions of the firm: considering its head office alone or considering the spatial environment of the majority of its employees. The following section presents the data and the methodology used.

DATA AND METHODOLOGY

This section presents the firm dataset linked to the industrial sector and the methodology used to build the indices of specialization and diversification. These indices will help to build the variables related to the geographic environment of the firm with respect to two conceptions: one reduced to the head office of the firm and another considering a larger conception of the firm. These spatial environments of the firms will be introduced in the econometric estimations presented in the fourth section.

Dataset

Three mandatory public surveys conducted by the INSEE were used. The CIS8 (Community Innovation Survey 2008) and the EAE 2007 (The Annual Survey of Firms of 2007) surveys provide the necessary data to construct a population of industrial firms that will serve as the reference population for the analysis of innovation. The DADS 2008 survey (Annual Declaration of Social Data of 2008), which is relevant to the unit level, is used for calculating the indices of specialization and diversification.

The CIS8 is a declarative sample survey that aims to evaluate firm performance in terms of innovation during a period of three years, i.e. from 2006 to 2008. It makes it possible to characterize the processes of innovation that occur within a firm, taking into account the influence of the characteristics of firms and inter-firm relationships, particularly in terms of the sources of information and of the cooperation toward innovation. This survey is combined with the 2007 Annual Survey of Firms, which also provides general information about the firms and their units. Thus, one has a database of 4703 firms of 20 employees or more, representative of the French industrial sector. Of the firms, 36.2% have created innovative products with an average turnover share in improved goods of 27.15% (for the descriptive statistics, see Table A1 in Appendix A).

Lastly, the DADS 2008 census provides information about all of the units of the manufacturing sector. The 106763 industrial units with at least one employee serve as a basis for calculating the specialization and diversification indices per employment area used for characterizing the spatial profile of our population of firms.

Specialization and diversification indicators

In keeping with the majority of the studies on the subject, the most common indicators for measuring the externalities of specialization and diversification have been used, i.e. LQ and Herfindahl Index. As these indices are highly sensitive to how the divisions into geographic areas and sectors are accomplished, the choice of divisions is therefore presented.

Geographic zoning and nomenclature of the industrial activities. The spatial entity selected is 'employment zone',[2] which is similar to the 'travel to work area' in Great Britain or to the 'Labour market areas' in the United States. There are 304 employment zones in metropolitan France. The division into sectors is based on the Nomenclature of French Activities (NAF rev. 2) and the associated Aggregated Nomenclature (NA 2008). It was chosen to divide the industrial sector into 12 large categories (two-digit) of activities in the manufacturing industry. Other more specific divisions were available, but they were not suitable because certain sectors became too detailed (resulting in a low number of units) while others did not.[3] This breakdown is applied to all industrial units (DADS data) of more than one employee located in metropolitan France (Table A2 in Appendix A). The remainder of this article will use the term 'sector of activities' to refer to these categories of activities.

The specialization index. The LQ is the most commonly used indicator in the literature for the evaluation of the specialization externalities (i.e., MAR externalities) (O'Donoghue and Gleave, 2004; Beaudry and Schiffauerova, 2009). It is calculated as the weight of a given sector k in a given geographic area i, taking into account the total weight of the sector k in the entire territory.

$$LQ_{k,i} = \frac{s_{k,i}}{s_k}$$

An LQ above 1 indicates that sector k is over-represented in zone i relative to the nation as a whole, which indicates a relative specialization of the zone. However, it may appear arbitrary to choose such a threshold to identify the specialization externalities linked to the existence of clusters of firms. As to this day there has been no theoretical arguments justifying a division threshold, a statistical approach was chosen by selecting as the specialization threshold the third quartile of the distribution of the LQs for all sectors. The value of the third quartile for all 3313 calculated LQ is equal to 1.3672.[4] By applying this threshold value, 828 specialized 'areas/sectors' were identified, which are called 'clusters' in the remainder of this article. They represent 31.3% of the units and 49.9% of the employees working in the sector. The number of specialized sectors varies according to the zones and sectors of activities (for the descriptive statistics, see Fig. A1 in Appendix A). According to this threshold value, four sectors of activities present an over-representation of the specialized sectors: agribusiness (CA), the wood and paper manufacturing sector (CC), the pharmaceutical sector (CF), the manufacture of rubber and plastics products, and other non-metallic mineral products (CG).

Index of diversification. To evaluate the level of diversification in the employment zones, the Herfindahl Index was applied, which is used extensively in the literature and makes it possible to compare the distribution of the workforces in each zone according to the type of sector classification adopted. The index maximal value 1 indicates that all employment is

Table 1. Distribution of specialized sectors according to the diversification of employment areas

Number of specialized areas/sectors or clusters per area	Diversified employment area	Not diversified employment area	Total employment area
0	0	0	0
1	3	23	26
2	18	78	96
3	33	91	124
4	19	33	52
5	3	3	6
Total	76	228	304

Source: DADS 2008, INSEE.

concentrated in one sector. It is therefore the low values of the index that indicate a diversity of activities that is likely to foster diversification externalities. The Herfindahl index for an area i with n sectors is written as follows:

$$H_i = \sum_{k=1}^{n} (s_{k,i})^2$$

Here again, there are no theoretical criteria that could be used to fix a threshold of determination of the diversification externalities. As for LQ, the statistical threshold selected is set at the first quartile. Thus, 25% of all zones present a diversified industrial employment structure, which corresponds to a Herfindahl index below 0.1418. This corresponds to 76 employment zones representing 49.6% of all the units and 46.4% of all the industrial employees.

With the help of both of these indices, one can evaluate whether an employment area is diversified or not and whether it has one or more specialized industrial sector(s). It is interesting to note that all 304 employment zones in the French territory have at least one specialized sector (Table 1). Furthermore, the specialized sectors situated in the 76 diversified zones present a little over one-quarter of all of the specialized sectors. The combined effect therefore represents one-quarter of the zones; and three-quarters of the zones are not diversified (and only specialized). It was also test whether the size of the area influences the probability of being a diversified area or of having a greater or lesser number of specialized area/sectors. The results indicate that diversified areas tend to be more frequent for larger areas; however, there is no significant relationship between the size of the areas and their specialization.[5]

The combined presence of diversified and specialized zones leads one to investigate the problem of constructing indicators that describe the geographic environment of the firms surveyed for the purpose of evaluating their impacts on the innovation performance.

Indicators of the firm's spatial environment

The authors wanted to build indicators of the firm geographic environment that take into account the fact that a firm can have several units and can therefore be located in several types of areas. As these areas could be specialized and/or diversified according to what is exposed below, the locations of the units of the firm lead one to define the spatial profiles of the firms. To build these profiles, the data on firm units obtained from the 2007 Annual Survey on Firms are used. Thus, one has a population of 20 405 industrial firms of at least 20 employees and distributed across 42 921 units.

Measure at the unit level. Each unit u can be classified according to the employment zone in which it is situated and to the sector of activity to which the firm belongs:[6]

For the specialisation area location: spe_u

$$= \begin{cases} 0 \text{ if } u \text{ is in non-specialized area} \\ \quad \text{(according to firm's activity sector)} \\ 1 \text{ if } u \text{ is in specialized area} \\ \quad \text{(according to firm's activity sector)} \end{cases}$$

For the diversification area location: div_u

$$= \begin{cases} 0 \text{ if } u \text{ is in non-diversified area} \\ 1 \text{ if } u \text{ is in diversified area} \end{cases}$$

One can then define a mode of calculation of the spatial profiles for the firms that possess several units, bearing in mind the fact that various profiles can be found. All the units of a firm can be located in the same employment zone. The units of a firm can also be located in several employment zones with the same specialization or diversification characteristics. Finally, the units of a firm can be located in several employment zones with opposing characteristics in terms of the specialization or diversification. The latter case applies to 39% of the multi-unit firms for specialization and to 45% for diversification. As it seems difficult to test for the externalities in the case of firms that have units in different types of zones in terms of specialization and/or diversification, two modes of calculation were selected that enable one to allocate a multi-unit firm to a single type of zone in terms of specialization or diversification.

Measure at the firm level. The first calculation concerns the firm's head office. It is based on the hypothesis that the head office is the focal point of the inter-firm relations. Thus, the spatial profiles of the firm for specialization and diversification correspond to that of its head office. In this case, the unit considered is the head office, that denotes $u = H$.

Table 2. Distribution of firms according to their spatial profiles

Firm's environment based on the majority of employees for multi-unit firm	Number of firms		
	Count data	Weighted count data	Per cent of firms
Specialized environment			
Single-unit firm in non-specialized area	1528	7534	40.1
Single-unit firm in specialized area	1000	4399	23.4
Total for single-unit firm	2528	11 933	63.5
Multi-unit firm in non-specialized area	1126	3870	20.6
Multi-unit firm in specialized area	1049	2999	16.0
Total for multi-unit firm	2175	6870	36.5
Total	4703	18 803	100.0
Diversified environment			
Single-unit firm in non-diversified area	1348	6311	33.6
Single-unit firm in diversified area	1180	5622	29.9
Total for single-unit firm	2528	11 933	63.5
Multi-unit firm in non-diversified area	1143	3541	18.8
Multi-unit firm in diversified area	1032	3329	17.7
Total for multi-unit firm	2175	6870	36.5
Total	4703	18 803	100.0
Specialized and diversified environment			
Single-unit in specialized and diversified area	366	1661	8.8
Single-unit in specialized and non-diversified area	634	2739	14.6
Single-unit in non-special. area	1528	7534	40.1
Total for single-unit firm	2528	11 933	63.5
Multi-units in specialized and diversified area	406	1165	6.2
Multi-units in specialized and non-diversified area	643	1834	9.8
Multi-units in non-specialized area	1126	3870	20.6
Total for multi-unit firm	2175	6870	36.5
Total	4703	18 803	100.0

Note: The representativeness of these spatial profiles was tested by comparing them with spatial profiles calculated on exhaustive firm's data from EAE 2007.
Sources: DADS 2008, CIS8, INSEE.

However, this first approach seems unsatisfactory with regard to the question of spatial externalities in so far as the head office can be an administrative unit, and the industrial plants of the firm can be in other employment areas. For this second calculation, a criterion was therefore used whereby at least 50% of the firm's total workforce is located in zones that are likely to generate the same type of externalities. Thus, the spatial profiles for firms take into account the spatial environment of each unit *u* of the firm associated with the number of employees *e*:

$$\text{spe}_{50} = \begin{cases} 0 \text{ if } \sum_u (\text{e}_u \times \text{spe}_u) < \left(\frac{\text{e}_f}{2}\right) \\ 1 \text{ if } \sum_u (\text{e}_u \times \text{spe}_u) < \left(\frac{\text{e}_f}{2}\right) \end{cases}$$

$$\text{div}_{50} = \begin{cases} 0 \text{ if } \sum_u (\text{e}_u \times \text{div}_u) < \left(\frac{\text{e}_f}{2}\right) \\ 1 \text{ if } \sum_u (\text{e}_u \times \text{div}_u) < \left(\frac{\text{e}_f}{2}\right) \end{cases}$$

On this basis, one will be able to build different types of spatial profiles for the firms, taking into account the degree of specialization and the diversification through the spatial organizational structure of the firm that differs if the firm is a single- or multi-unit firm. These two modes of calculation are applied to the firm data (CIS8) (see Table 2 for the results based on the criteria of at least 50% of the total firm workforce).

ECONOMETRIC ESTIMATIONS AND RESULTS

The selection model used to characterize both stages of the firms' decision to engage in innovation activities and that concerns the intensity of the innovation activities is first presented. Then the influence of the two spatial profiles (the head office versus the majority of the employees) defined above on the firm's decision to innovate is compared. Next, the effects of specialization and diversification on both stages of innovation (engagement and intensity) using the selection model are tested.

The selection model

An innovating firm can be analysed as the result of a twofold decision making process: the decision to innovate or not, and if the firm does innovate, the decision as to what extent of the firm's activities should be dedicated to innovation. The estimation of the second equation only concerns the firms that have innovated ($I = 1$). Thus, one faces a selection bias ('incidental truncation'; WOOLDRIDGE, 2002) that affects the estimation of the regressors β.

With regard to the first decision, innovating requires that the firm invest in innovation activities. This preliminary decision is generally based on the difference between the benefits expected from the innovation activities and their costs. This unobserved difference refers to a latent variable I^* such that:

$$I_i^* = \gamma Z_i + \varepsilon_i$$

with ε_i the error term of the distribution N(0,1); where Z is a vector of independent variables that characterize the spatial environment of the firm that could be diversified (Div) and/or specialized (Spe) combined with the organizational structure of the firm that could be a single- or multi-unit firm (S): $C_i(\text{Spe}, \text{Div}, S)$. The equation could contain other control variables relative to other firm characteristics (FC), to the sectoral and market environment (E), and to the innovation profile (IP).

The equation could be written according to the spatial profile indicators retained:

$$I_i^* = \gamma_0 + \gamma_1 C_i(\text{Spe}, \text{Div}, S) + \gamma_2 \text{FC}_i + \gamma_3 E_i + \gamma_4 \text{IP}_i + \varepsilon_i$$

If $I^* > 0$, then the firm has innovated (i.e., its innovation activities have led to the firm selling new products): note that I is the observed variable indicating whether the firm has innovated or not (i.e., has introduced a new product on the market[7]):

$$\begin{cases} I_i = 1 \text{ if } I_i^* > 0 \\ I_i = 0 \text{ if } I_i^* \leq 0 \end{cases}$$

The probability that firm i has innovated is:

$$\begin{aligned} P(I_i = 1) &= P(I_i^* > 0) \\ &= P(\varepsilon_i > -\gamma_0 - \gamma_1 C_i(\text{Spe}, \text{Div}, S) \\ &\quad - \gamma_2 \text{FC}_i - \gamma_3 E_i - \gamma_4 \text{IP}) \end{aligned}$$

$$\begin{aligned} P(I_i = 1) &= P(I_i^* > 0) \\ &= \Phi(\gamma_0 + \gamma_1 C_i(\text{Spe}, \text{Div}, S) + \gamma_2 \text{FC}_i + \gamma_3 E_i \\ &\quad + \gamma_4 \text{IP}_i) \end{aligned} \quad (1)$$

with Φ as the standard distribution function. This probability is estimated using a probit model that enables one to evaluate the weight of the independent variables that have an influence on the propensity of the firm to commercialize new products.

The second decision is related to the innovation intensity. This variable is constructed on the basis of the logit[8] of the percentage of the turnover attributable to the sale of new product(s) (MAIRESSE and MOHNEN, 2010):

$$y_i = \frac{\text{turnover due to new products}}{\text{turnover}}$$

$$\text{i.e.} \quad Y_i = \text{Logit}(y_i) = \frac{1 - y_i}{y_i}$$

This innovation intensity depends on a number of independent variables, among which a certain number can be common to those that explain the decision to innovate or not and can belong to the same groups of control variables (FC, E, IP).

$$\begin{aligned} Y_i = {} & \beta_0 + \beta_1 C_i(\text{Spe}, \text{Div}, S) + \beta_2 \text{FC}_i + \beta_3 E_i \\ & + \beta_4 \text{IP}_i + u_i \end{aligned} \quad (2)$$

with u_i as the term of error of standard distribution.

There likely are unobservable factors, which, via the observed variables X, influence both the probability of the firm to innovate and the innovation intensity. Thus, the correlation between the residuals of the two equations skews the estimations ($\text{cov}(u, \varepsilon) \neq 0$). To control for this bias, HECKMAN's (1979) maximum likelihood procedure is used, in which the probability to innovate and the intensity of this innovation are jointly estimated while controlling for the correlation between the unobservable variables. It is assumed that the error terms u_i and ε_i follow a bivariate normal distribution with mean 0 and correlation function ρ:

$$(u, \varepsilon) \rightarrow N\begin{pmatrix} 0 & \sigma_u & \rho \\ 0 & \rho & \sigma_\varepsilon \end{pmatrix}$$

According to the hypothesis of residual normality, it is demonstrated that the regression equation is:

$$\begin{aligned} (Y_i | I_i = 1) \\ = \beta_0 + \beta_1 C_i(\text{Spe}, \text{Div}, S) + \beta_2 \text{FC}_i + \beta_3 E_i \\ + \beta_4 \text{IP}_i + \rho \sigma_u \lambda_i + v_i \end{aligned}$$

where:

$$\lambda_i = \frac{\phi(-\gamma_0 - \gamma_1 C_i(\text{Spe}, \text{Div}, S) - \gamma_2 \text{FC}_i - \gamma_3 E_i - \gamma_4 \text{IP}/\sigma_\varepsilon)}{[1 - \Phi(-\gamma_0 - \gamma_1 C_i(\text{Spe}, \text{Div}, S) - \gamma_2 \text{FC}_i - \gamma_3 E_i - \gamma_4 \text{IP}/\sigma_\varepsilon)]}$$

corresponds to the inverse of the Mills ratio, a term that corrects the selectivity bias in the estimation of Y; and where $\rho\sigma_u$ is considered as the regression coefficient of the inverse of the Mills ratio. The Heckman procedure enables one to perform an unbiased estimation of the regression coefficients for the non-random sample of innovating firms.

Spatial environment and the decision to innovate: the firm's head office analysis versus the firm's units analysis

Equation (1) is used to test the influence of the mode of calculation of the 'firm's geographic environment' variable (based on the head office or 50% of the firm's total workforce) on the decision to innovate.[9] This involves evaluating the impact of the different types of environment on innovation and specifying the value added with the measurement of the firm units. The difficulty lies in the possible existence of combined specialization and diversification externalities. Furthermore, it is important to use the firm organizational structure, i.e. the fact that it is a single-unit firm or a multi-unit organization, as a control variable.

The effects of specialization and diversification on innovation are tested by including the effect of the organizational structure. Much of the literature points to the difficulties of assessing the effects of the interactions in the case of the non-linear models (AI and NORTON, 2003). The marginal effects must be calculated by the difference and not by the derivation. BUIS (2010) proposes an alternative solution based on the odds ratios that make it possible to measure the multiplicative effect of an interaction. Note that this ratio measures the growth or decline of the dependent variable for a change of 1 unit of the independent variable. To compare the models, it is also important to identify the effect of the dependent variable, i.e. the decision to innovate, all things being equal. This is possible by building a binary variable that will take the value 1 for the whole population.

First, the interaction between specialization and diversification is tested for both of the modes of calculation of the indicators (Models 1 and 2); specialization, diversification and organizational structure are then combined (Models 3 and 4). These models are specified without considering the other control variables mentioned above. The general specification of the equation (1) estimated is:

$$P(I_i = 1) = \Phi(\gamma_0 + \gamma_1 C_i(\text{Spe, Div}, S))$$

Next, four models are estimated and presented in Table 3, depending on the combination of the variables characterizing the spatial profile of the firm.

The main results indicate a small but significant difference between the two modes of the calculation of the variable 'firm's geographic environment' and a significant effect of the interaction between specialization and diversification. In addition, being a multi-unit firm has a strongly positive effect on product innovation. More precisely, taking into account the interaction between diversification and specialization does not reveal any significant difference in terms of the way the variable 'firm's geographic environment' is calculated. The results of Models 1 and 2 are relatively similar, except for the fact that the effect of diversification on innovation seems to have a more positive influence when the calculation involves the head office environment as opposed to 50% of the firm's workforce. Moreover, the most significant result of these two models is the positive effect of the interaction between diversification and specialization on the innovation performances for the firms that combine the benefits of being located in a specialized and diversified environment. Thus, Model 1 indicates that the firms located in both the specialized and diversified zones have the highest probability of innovating (they are 26.7% more likely to innovate than firms located in another type of geographic environment). This joint effect proves more efficient than the isolated effect of being located in a zone that is either diversified or specialized.

In Models 3 and 4, adding the firm organizational structure to the specialization and diversification criterion reveals the impact of the former on the innovation performance. Thus, being a multi-unit firm strongly impacts the propensity of the firm to innovate, regardless of its geographic environment. This effect is stable; indeed, as shown in the following model, it is also observed when the traditional determinants of innovation (in particular the size) are controlled for. Furthermore, the effect of the simultaneous presence of specialization and diversification is always positive and significant in these models (3 and 4). And the positive effect of this interaction is even more important when the firm is a multi-unit organization. Thus, Model 4 indicates that the probability of a firm to innovate is 1.26 times higher when it combines a multi-unit structure with a presence in both the diversified and specialized types of environments. Finally, it is observed that being a multi-unit firm located in a diversified zone has a negative effect on its probability to innovate. This result is significant in Model 4, whereas it is not in Model 3. This confirms that the impact of the firm's geographic environment on its propensity to innovate varies depending on whether the analysis takes into account the location of the head office or all of the firm units. The effect of diversification on the choice is more dependent on the head office than the secondary units. Similarly, in the analysis based on 50% of the workforce, the diversified environment benefits the single firm more than the multi-unit firm; more precisely, it appears that the single firm is more dependent on its environment in the step of engaging in innovation than a multi-establishment firm that has more sources of spillovers. This tends to indicate that complexity can be taken into account if in the analysis one takes into

Table 3. Influence of the firm's spatial environment (specialization and/or diversification area) and of the firm's organizational structure on its innovation, based on two approaches (on the head office or on the firm's workforce)

Dependant variable: Firm has innovated in products	(1) Head office approach	(2) 50% employees approach	(3) Head office approach	(4) 50% employees approach
Specialized area	1.050	1.060	1.039	1.039
	(1.11)	(1.38)	(0.70)	(0.70)
Diversified area	1.052	0.982	1.047	1.047
	(1.28)	(−0.47)	(0.92)	(0.92)
Multi-unit firm			1.628***	1.710***
			(7.72)	(8.82)
Double interaction				
Specialized and diversified area	1.267***	1.289***	1.196**	1.196**
	(3.76)	(4.03)	(2.17)	(2.17)
Multi-unit firm and specialized area			0.947	0.950
			(−0.60)	(−0.59)
Multi-unit firm and diversified area			0.912	0.808***
			(−1.12)	(−2.59)
Triple interaction				
Specialized and diversified area for a multi-unit firm			1.189	1.262*
			(1.33)	(1.80)
Baseline	0.522***	0.537***	0.448***	0.448***
	(−22.05)	(−21.54)	(−22.20)	(−22.20)
chi2	1445.91	1440.98	1607.84	1611.31
P	0.000	0.000	0.000	0.000
Akaike information criterion (AIC)	24 560.5	24 566.3	24 364.9	24 360.9
N	4703	4703	4703	4703

Notes: Exponentiated coefficients – $*p < 0.10$, $**p < 0.05$, $***p < 0.01$ – z-values are given in parentheses.

Baseline: in model 1, for firms located in areas that are neither diversified nor specialized, there are approximately two firms that do not innovate for one that does (baseline = 0.522).

Sources: DADS 2008, EAE 2007, CIS8, INSEE.

account the units of the firm as opposed to its head office only.

In view of the results and to obtain a more rigorous evaluation of the effects of the externalities on the innovation performance of the firm, a method of analysis is selected that takes into consideration the location of the majority of the firm's employees. This more global approach to the organization enables one to obtain a more complete view of the externalities that operate not just at the head office level but also at the level of the various units of the multi-unit firms.[10]

Differentiated effects of spatial externalities on innovation

The goal of this section is to test the role of the types of spatial externalities (MAR, Jacobs and both sets) on both stages of the decision-making process related to innovation: the decision to engage in innovation activities and the decision concerning the intensity of innovation. The spatial profiles take into account the firm organizational structure (single- or multi-unit), and the model includes the other determinants of the propensity of the firm to innovate, related to their structural characteristics, their relationship with external actors and their sectoral and competitive environments. The latter variables function here as control variables, according to the equations presented in the fourth section. Three models are tested to account for specialization, diversification and both sets (Table 4).

The specialization model (M1), which pertains to the Marshallian externalities, highlights the absence of the influence of the specialization area on the single-unit firms' engagement in innovation activities. On the other hand, multi-unit firms have a higher propensity to engage in innovation activities, and it is all the higher when they are located in specialized zones. As for the innovation intensity, the effect of the zone specialization is much more significant. It can be observed that being located in a specialized environment has a positive effect on the propensity of a firm to innovate and that this effect is stronger for multi-unit firms than for single-unit firms. Simultaneously, it is found that being located in non-specialized zones has no effect on the innovation propensity of multi-unit firms. Thus, the first model tends to indicate, first, that the form of organization does have an influence on a firm's engagement in innovation activities and that the single-unit firms and multi-unit firms do not have the

Table 4. Effects of specialized, diversified and dual areas for single- or multi-unit firms in Heckman's models

Step 1: Product innovation no/yes	M1: Spe50		M2: Div50		M3: Spe50 and Div50	
Step 2: Logit of turnover share in improved goods	S1: Choice	S2: Intensity	S1: Choice	S2: Intensity	S1: Choice	S2: Intensity
I. Spatial profiles						
Single unit and not specialized area	Reference	Reference				
Single and specialized	0.040	0.373***				
Multi- and specialized	0.132***	0.477***				
Multi- and not specialized	0.107***	0.059				
Single unit and not diversified area			Reference	Reference		
Single and diversified			0.071***	−0.260***		
Multi- and diversified			0.149***	0.048		
Multi- and not diversified			0.125***	−0.074		
Single unit and not specialized area					Reference	Reference
Single, specialized and diversified					0.083**	0.028
Single specialized and not diversified					0.013	0.611***
Multi- and not specialized					0.108***	0.048
Multi, specialized and diversified					0.180***	0.318**
Multi, specialized and not diversified					0.103**	0.572***
Characteristics of the firm						
Size: 250 employees and more	Reference	Reference	Reference	Reference	Reference	Reference
Between 50 and 249 employees	−0.311***	0.410***	−0.314***	0.341***	−0.312***	0.416***
Fewer than 50 employees	−0.545***	0.615***	−0.551***	0.546***	−0.546***	0.613***
Part of a group	0.186***	−0.125**	0.186***	−0.110*	0.185***	−0.116*
Number of units (in log)	0.008	−0.089	0.007	−0.105	0.008	−0.081
Sectoral and market environment						
Sector: Agro-food activities	Reference	Reference	Reference	Reference	Reference	Reference
Intermediates activities	0.013	0.687***	0.003	0.673***	0.009	0.713***
Transport activities	−0.256***	0.798***	−0.262***	0.772***	−0.255***	0.776***
Equipment activities	0.297***	0.553***	0.285***	0.581***	0.285***	0.635***
Consumer-related activities	−0.108***	0.387***	−0.119***	0.375***	−0.117***	0.440***
Market concentration	0.792***	−0.292	0.791***	−0.197	0.780***	−0.221
Market: national largest market	Reference		Reference		Reference	
Local/regional largest market	−0.466***		−0.468***		−0.466***	
Other countries largest market	0.213***		0.215***		0.212***	
Innovative profile						
Share of R&D expenses in turnover		4.273***		4.468***		4.349***
Internal sources of information		−0.054		−0.054		−0.060
Suppliers sources of information		0.192***		0.199***		0.184***
Clients sources of information		0.192***		0.173***		0.181***
Consultants sources of information		0.239**		0.279**		0.263**

(Continued)

Table 4. Continued

Step 1: Product innovation no/yes Step 2: Logit of turnover share in improved goods	M1: Spe50 S1: Choice	M1: Spe50 S2: Intensity	M2: Div50 S1: Choice	M2: Div50 S2: Intensity	M3: Spe50 and Div50 S1: Choice	M3: Spe50 and Div50 S2: Intensity
Cooperation inside group		0.154**		0.159**		0.149**
Cooperation with supplier		−0.155**		−0.174**		−0.169**
Cooperation with client		0.019		0.021		0.021
Cooperation with competitor		0.007		0.042		0.023
Cooperation with consultant		−0.224***		−0.246***		−0.220***
Cooperation with university		−0.228***		−0.237***		−0.217**
Cooperation with public research institute		0.518***		0.542***		0.502***
New practices for organizing procedures	0.519***		0.518***		0.519***	
New methods of organizing work responsibilities	0.220***		0.218***		0.220***	
New methods of organizing external relations	0.262***		0.264***		0.262***	
Constant	−0.322***	−2.075***	−0.326***	−1.777***	−0.315***	−2.120***
athrho	−0.169***		−0.154***		−0.167***	−0.167***
ln sigma	0.753***		0.754***		0.750***	0.750***
chi2	323.588		281.776		354.793	
p	0.000		0.000		0.000	
Akaike information criterion (AIC)	50 317.71		50 354.75		50 291.97	
N	4703		4703		4703	
*N*_cens	2624		2624		2624	

Note: $*p < 0.10$, $**p < 0.05$, $***p < 0.01$.
Sources: EAE 2007, DADS 2008, CIS8, INSEE.

same relation to space; and, second, that the Marshallian externalities have a particularly strong impact on the firm innovation intensity.

In the diversification (Jacobian) model (M2), the results highlight different effects depending on the type of firm and on the stage of the innovation process. Thus, though the engagement in innovation activities of single-unit firms is positively influenced by being located in a diversified environment, their presence in this type of environment has a highly significant and negative effect on the intensity of innovation performance. In the case of multi-unit firms, being located in diversified zones increases the probability of innovating but has no effect on intensity. Thus, the second model tends to indicate that the Jacobian externalities have a particularly strong impact on and are necessary for the single-unit firm's decision to innovate but that, conversely, they have a negative impact on innovation intensity.

Model 3 tests the influence of the combination of specialization and diversification.[11] For the first stage of the decision-making process (to engage or not to engage in innovation activities), it can be observed that being located in a specialized environment has a positive influence on innovation when it is combined with being located in a diversified zone, bearing in mind that a multi-unit firm always has a higher propensity to innovate than a single-unit firm, regardless of the geographic environment. As far as innovation intensity is concerned, the result is original in that it demonstrates that the innovation intensity is higher for firms that operate in a specialized but non-diversified environment. Here, the idea is confirmed that the Jacobian externalities can have a negative impact on the firm innovation intensity. This is particularly marked for single-unit firms, which are more dependent upon a single location.

In brief, the results are complementary and corroborate one another. One can observe a favourable influence of being a multi-unit organization on the decision to innovate and find that the two types of organizations (multi- versus single-unit) have noticeably different relations to space. For single-unit firms, being located in a diversified environment that generates various externalities and knowledge spillovers has a positive impact on the decision to engage in innovation. However, the intensity of the innovation performance depends more on operating in a Marshallian environment and on exchanges of knowledge related to firm-specific activities. The innovation intensity of single-unit firms is negatively impacted by a diversified environment. Multi-unit firms are less dependent on the characteristics of the environment but being located in a diversified environment does have an effect on the decision to engage in innovation: Specialization appears as a second-order factor that has a more positive effect when it is combined with diversification. Inversely, the innovation intensity of a multi-unit firm is positively influenced by the fact that its environment is specialized and negatively influenced by its being diversified.

As for the control variables, the effects are similar for the three models and are in keeping with what is generally emphasized in the literature. However, our results concerning the second stage of the decision-making process, i.e. concerning the intensity of the innovation activities, are original.

Thus, the decision to innovate is positively influenced by the size of the firm, by the fact that it belongs to a group and by its openness to international markets. The joint implementation of product and organizational innovations, particularly in terms of procedures, also positively influences the propensity to innovate. Finally, the market concentration, which is revealing of the degree of competitive pressure, plays a positive role in the decision to innovate. In comparison with the other sectors, the agribusiness firms innovate more than the firms in the transport and intermediate goods industries but less than the firms in the equipment industries. Finally, the number of units per firm has no effect in the models.

The results concerning the effect of the firm sector and size on the intensity of innovation activities are in keeping with those in the literature. The size has a negative effect that can be explained by the fact that the smaller the firm, the higher the probability that a high percentage of its turnover is attributable to the sale of new products. It has also been found that belonging to the agribusiness sector has a negative effect, which can be explained by the low level of radical innovation for the agro-food firms and therefore by the fact that they never renew their product ranges, an activity that leads to high levels of innovation intensity. Belonging to a group and market concentration also have a negative influence, though the influence of market concentration is non-significant. Another classic result pertains to the close relation between innovation intensity and R&D expenditures. Finally, contrasting results about the sources of information and innovation cooperation are obtained. The market information sources play a positive role in firm innovation intensity, whether they be downstream or upstream sources of information or information contributed by external consultants. Internal sources of information, within the firm or the group it belongs to, have a slightly negative though non-significant effect. However, innovation-driven cooperations reveal a different configuration. Cooperation with the other firms of the group has a positive effect on the innovation intensity, whereas cooperation with suppliers or consultants has a negative influence, and cooperation with clients has no significant effect. Finally, cooperation with research organizations has a positive influence on innovation intensity, whereas cooperation with universities has a negative influence.

DISCUSSION AND CONCLUSIONS

The aim of this study has been to analyse the influence on innovation performance of the spatial externalities related to the location of firms. The idea has been to take into account, beyond the Marshall–Jacobs dichotomy, the complexity of the various types of environment and the possible combination of both types of externalities in the same area (dual model). The originality of this study also lies in the use of a larger definition of the firm, i.e. one that is not limited to the head office, to evaluate better the impact of the spatial externalities on innovation. This study used a database of French industrial firms and a methodology that has enabled the authors to characterize the industrial content (the types of local clusters and indicators of diversity) of all the employment zones of the French territory.

Different results can be highlighted. The first is methodological. The analysis of the effects of the firm spatial environment on innovation performance has different results depending on whether one takes into consideration head offices or the organization as a whole (the calculation of the entire workforce of the firm). The difference is all the more significant as multi-unit firms – which can potentially be located in a variety of areas – play an important role in the productive system observed.

With regard to the Marshall–Jacobs dichotomy, the originality of this study lies in having tested the impact of both types of externalities separately and the impact of the combination of both. Overall, the results tend to indicate that the geographic environment of a firm does not have the same effect on the decision to engage in innovation activities as does the intensity of the innovation activities and that the effects also vary depending on the firm organizational structure. Furthermore, the diversification model tends to reveal the stimulating importance of the 'Jacobian' diversity in the decision of firms to engage in innovation and its negative effect on the innovation intensity. This is particularly marked in the case of single-unit firms, which are confined to one location. Their presence in a diversified area has in addition a negative effect on the intensity of the innovation activities. The specialization model indicates that being located in the specialized zones has, on the contrary, a more positive influence on the decision to innovate and reinforces innovation intensity.

The estimated dual model enables one to take into account the fact that an area can be both diversified (with a large number of activities) and specialized in the activity in which the firm in question is engaged. The results demonstrate that diversification combined with specialization increases the probability of firms to innovate and may be a sine qua non to innovation (being located in a zone that is specialized and non-diversified has little or no effect on the decision to innovate). The Jacobian externalities therefore play a particularly positive role in this stage of the decision-making process. On the other hand, diversification negatively influences the intensity of innovation that could be more dependent upon the exchanges of specialized knowledge. In this respect, one can observe once more the importance of the Marshallian effects and of the cluster processes in the intensity of the innovation processes.

Finally, the internal characteristics of firms have different effects depending on the stage of the innovation process, but these effects are similar for the three models. The decision of a firm to engage in innovation activities is particularly sensitive to its capacity to mobilize internal resources (e.g., large size and belonging to a group) and to the stimuli from the market (international openness, competitive pressure). Innovation intensity is influenced by the firm R&D expenditures, by its small size and by a strategy consisting of exchanging information with clients and suppliers. Cooperative relations, by contrast, tend to develop internally within the group and with public research organizations, which seems to indicate a search for trust-based and non-competitive relationships. These effects are not neutral in the firms' relation to space. They often indicate needs for organizational or geographical proximity – necessary for innovation – which are at the basis of the spatial cluster effects. The stability of the effects of the control variables revealed by the three models highlights all the more the specific effects of the spatial environment variables. This finding prompts one to conclude with the potential benefits of a more in-depth theoretical and methodological analysis to obtain a better understanding of the impact of spatial externalities on firm innovation behaviour, taking into account the possible choices of multi-location. Indeed, a deeper definition of the firm may change the understanding of the innovation processes for multi-unit firms. From this perspective, it would be important to test the robustness of the results by considering various definitions of the multi-unit firm, including one relative to its functional structure.

Acknowledgments – This work was carried out as part of the LABEX SMS [reference number ANR-11-LABX-0066] and also received financial support from the Institut National de la Recherche Agronomique (INRA) and the Regional Council of Midi-Pyrénées as part of PSDR project 'Compter'.

Funding – The authors would like to thank the editors and two anonymous reviewers for helpful comments.

APPENDIX A

Table A1. Descriptive statistics of the population of firms

Mean value of variables	All firms	Innovative firms in products
Product innovation	0.362	1
Turnover share in improved goods		27.15
Single firm	0.635	0.566
Multi-unit firms	0.365	0.434
Fewer than 50 employees	0.568	0.418
Between 50 and 249 employees	0.340	0.414
250 employees and more	0.092	0.169
Part of a group	0.505	0.637
New practices for organizing procedures	0.304	0.489
New methods of organizing work responsibilities	0.304	0.462
New methods of organizing external relations	0.143	0.242
Share of research and development (R&D) expenses in turnover		0.022
Internal sources of information		0.678
Suppliers sources of information		0.165
Clients sources of information		0.339
Cooperation inside group		0.284
Cooperation with supplier		0.323
Cooperation with client		0.283
Number of firms	4703	2079
Weighted number of firms	18 802	6800

Source: DADS 2008.

Table A2. Units and employees according to 'A38' manufacturing industrial activities

A38	Division	Title	Number of units	Number of employees	Average number of employees per unit
CA	10–12	Food and beverage	10 267	391 696	38.2
CB	13–15	Textile, wearing and leather	7307	137 498	18.8
CC	16–18	Wood, paper and printing	14 534	235 423	16.2
CE	20	Chemical	2838	152 142	53.6
CF	21	Pharmaceutical	651	85 757	131.7
CG	22–23	Rubber, plastic, non metallic	11 453	327 464	28.6
CH	24–25	Metal	17 250	447 293	25.9
CI	26	Computer, electronic, optic	3219	152 002	47.2
CJ	27	Electrical equipment	2442	130 852	53.6
CK	28	Machinery and equipment	6802	216 460	31.8
CL	29–30	Transport equipment	2844	391 859	137.8
CM	31–33	Furniture, other manufacturing	28 066	309 316	11.0
Total industry			107 673	2 977 762	27.7

Note: The category CD 'Coke and refined petroleum', which is made up 12353 employees in 105 units, is not taken into account in this manufacturing industry analysis. The numbers of employees are calculated using full-time equivalents.
Source: DADS 2008, INSEE.

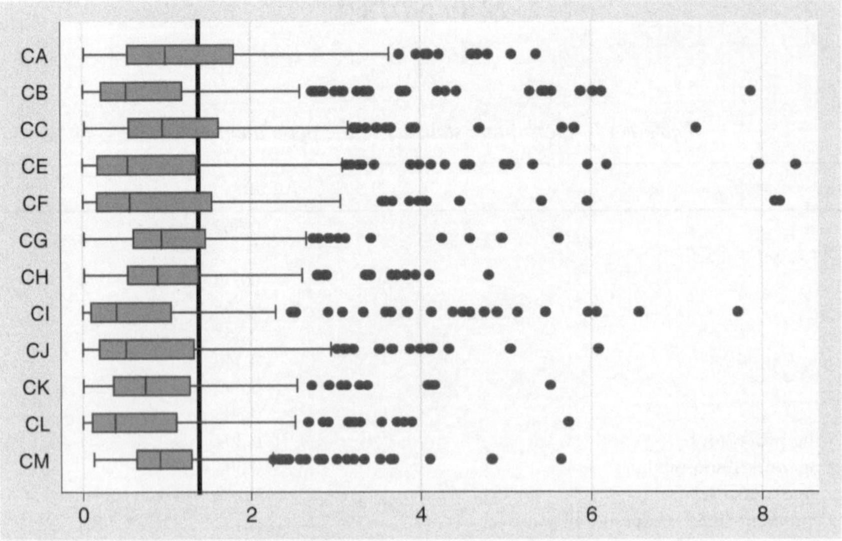

Fig. A1. *Distribution of location quotient (LQ) per industrial sector*
Note: The vertical bar shows the specialization cut-off for the LQ. Five extreme values of LQ > 10 are not represented.
Sources: DADS 2008, INSEE.

NOTES

1. As noted by BEAUDRY and SCHIFFEAUEROVA (2009), certain indicators of the Jacobs externalities do not measure diversity per se, but the size of the urbanization externality. These indicators capture instead the 'global urbanization externalities' related to market size but not the diversity implied by Jacobs externalities per se (measured by the Hirschman-Herfindahl index).
2. An employment zone is a geographical area within which most of the labour force lives and works and in which establishments can find the majority of the labour force necessary to occupy the offered jobs (INSEE definition). This zoning therefore has an economic meaning, unlike the divisions into administrative areas such as '*départements*'. The 2010 zoning is used here conducted on the basis of the 2006 home-to-work flows.
3. More precisely, level A38 of the 2008 NA is used, which is defined by INSEE as an international intermediate level between the sections and divisions, the two 'standard' levels that fit into the International (CITI rev. 4) and the European (NACE rev. 2) nomenclatures of activities. The 'division' level A88 will give 23 subsectors compared with the 12 subsectors with level A38, but the disaggregated sectors are, for example, part of the *Food and beverages industry* (divisions 10–12) or *Textile, wearing apparel and leather products sector* (divisions 13–15), which are not particularly suitable for the purpose of characterizing a breakdown into industrial sectors. See the description of the categories in the Table A2 in Appendix A.
4. As 304 employment zones and 12 sectors of activities are being worked with, i.e. 3648 possible combinations, it can be concluded that one sector is not represented in one zone for the 335 combinations.
5. The size of area is measured by the working population.
6. In this approach, all the units of the firm have for reference the main activity of the firm. This approach seems the simplest and most consistent to apply to an innovative firm because it allows taking into account all the units of the firm, including those with non-industrial activity. In addition, due to the organization into broad subsectors (two digits), the probability of having units in very different industrial sectors is relatively low.
7. New-to-firm or new-to-market products. The variable I includes 'radical' and 'incremental' product innovations.
8. Because the share of innovative sales is bounded by 0 and 1 (or by 0% and 100%), it is preferable to use the 'logit share' variable which can vary from $-\infty$ to $+\infty$ (MOHNEN *et al.*, 2006).
9. To interpret the results more simply, the choice here is to use the logistic distribution of the error term, which corresponds to a logit model.
10. As suggested by a referee, one may question the robustness of the choice of retaining the 50% threshold to affect multi-unit firms in a specialized environment or not (or in a diversified environment or not). To deal with this issue, 40% and 60% thresholds were tested. Approximately 3% of multi-unit firms see their environment to change with these new thresholds, compared with a 50% threshold. The results of the models (see the next section) were then compared with these different thresholds. The results appear robust, as shown by the significance of the explanatory variables which is not changed. This is supported by the small differences (< 0.01%) for the values of Akaike criterion in the estimates with different thresholds. Those estimations are available from the authors upon request.
11. Eight dummy variables are necessary to test all combinations of the diversity specialization and the type of firm. It was preferable to use an interpretation angle

with respect to the specialization externalities, not seeking to detail the externalities of diversification (or not) in the non-specialized areas. As the effect of diversification for single-unit firms in the non-specialized areas was not significant, the choice was not to distinguish diversified versus undiversified environment for this type of firm: the single-unit firms in non-specialized areas are taken as a reference for Model 3. And so on, the same was done for the multi-establishment firms. The estimates with eight dummies do not change the results (the detailed tables are available from the authors upon request).

REFERENCES

AI C. and NORTON E. (2003) Interaction terms in logit and probit models, *Economics Letters* **80**, 123–129. doi:10.1016/S0165-1765(03)00032-6

ANDERSEN E. and LUNDVALL B. A. (1988) Small national systems of innovation facing technological revolutions: an analytical framework, in FREEMAN C. and LUNDVALL B. (Eds) *Small Countries Facing the Technological Revolution*, pp. 9–36. Pinter, London.

AUDIA P. G., SORENSON O. and HAGE J. (2001) Tradeoffs in the organization of production: multiunit firms, geographic dispersion and organizational learning, in BAUM J. A. C. and GREVE H. R. (Eds) *Multiunit Organization and Multimarket Strategy, Advances in Strategic Management*, pp. 75–105. JAI Press, Oxford.

AUTANT-BERNARD C. and LESAGE J. P. (2011) Quantifying knowledge spillovers using spatial econometric models, *Journal of Regional Science* **51**, 471–496. doi:10.1111/j.1467-9787.2010.00705.x

ARROW K. J. (1962) The economic implications of learning by doing, *Review of Economic Studies* **29**, 155–172. doi:10.2307/2295952

AUDRETSCH D. B. and FELDMAN M. P. (2004) Knowledge spillovers and the geography of innovation, in HENDERSON J. V. and THISSE J. F. (Eds) *Handbook of Regional and Urban Economics*, vol. 4, pp. 2713–2739. Elsevier, Amsterdam.

BEARDSELL M. and HENDERSON V. (1999) Spatial evolution of the computer industry in the USA, *European Economic Review* **43**, 431–456. doi:10.1016/S0014-2921(98)00064-6

BEAUDRY C. and SCHIFFAUEROVA A. (2009) Who's right, Marshall or Jacobs? The localization versus urbanization debate, *Research Policy* **38**, 318–337. doi:10.1016/j.respol.2008.11.010

BOSCHMA R. (2005) Proximity and innovation: a critical assessment, *Regional Studies* **39**, 61–74. doi:10.1080/0034340052000320887

BUIS M. L. (2010) Interpretation of interactions in nonlinear models, *Stata Journal* **10**, 11–29.

COHEN W. and LEVINTHAL D. (1989) Innovation and learning: the two faces of R&D, *Economic Journal* **99**, 569–596. doi:10.2307/2233763

DICKEN P. and MALMBERG A. (2001) Firms in territories: a relational perspective, *Economic Geography* **77**, 345–363. doi:10.2307/3594105

FELDMAN M. P. and KOGLER D. F. (2010) Stylized facts in the geography of innovation, in BRONWYN H. and ROSENBERG N. (Eds) *Handbook of the Economics of Innovation*, vol. 1, pp. 382–410. Elsevier, Amsterdam.

FRENKEN K., VAN OORT F. G. and VERBURG T. (2007) Related variety, unrelated variety and regional economic growth, *Regional Studies* **41**, 685–697. doi:10.1080/00343400601120296

FU S. and HONG J. (2011) Testing urbanization economies in manufacturing industries: urban diversity or urban size?, *Journal of Regional Science* **51**, 585–603. doi:10.1111/j.1467-9787.2010.00702.x

GALLIANO D., ROUX P. and SOULIE N. (2011) ICT intensity of use and the geography of the firm, *Environment and Planning A* **43**, 67–86. doi:10.1068/a43167

HECKMAN J. J. (1979) Sample selection bias as a specification error, *Econometrica* **47**, 153–162. doi:10.2307/1912352

JACOBS J. (1969) *The Economies of Cities*, Random House, New York, NY.

GLAESER E., KALLAL H., SCHEINKMAN J. and SHLEIFER A. (1992) Growth in cities, *Journal of Political Economy* **100**, 1126–1152. doi:10.1086/261856

KARSHENAS M. and STONEMAN P. (1993) Rank, stock, order and epidemic effects in the diffusion of new process technologies: an empirical model, *RAND Journal of Economics* **24**, 503–528. doi:10.2307/2555742

LHUILLERY S. and PFISTER E. (2009) R&D cooperation and failures in innovation projects: empirical evidence from French CIS data, *Research Policy* **38**, 45–57. doi:10.1016/j.respol.2008.09.002

MAGRINI M. B. and GALLIANO D. (2012) Agglomeration economies, spatial organization and innovation performance, *Industry and Innovation* **19**, 607–630. doi:10.1080/13662716.2012.726809

MAIRESSE J. and MOHNEN P. (2010) Using innovation survey for econometric analysis, in BRONWYN H. and ROSENBERG N. (Eds) *Handbook of the Economics of Innovation*, vol. 2, pp. 1129–1155. Elsevier, Amsterdam.

MARSHALL A. (1890) *The Principles of Economics*. Macmillan, London.

MASSARD N. and MEHIER C. (2009) Proximity and innovation through an 'accessibility to knowledge' lens, *Regional studies* **43**, 77–88. doi:10.1080/00343400701808881

MCCANN B. T. and FOLTA T. B. (2011) Performance differentials within geographic clusters, *Journal of Business Venturing* **26**, 104–123. doi:10.1016/j.jbusvent.2009.04.004

MILGROM P. and ROBERTS J. (1990) The economics of modern manufacturing: technology, strategy and organisation, *American Economic Review* **80**, 511–528.

MOHNEN P., MAIRESSE J. and DAGENAIS M. (2006) Innovativity: a comparison across seven European countries, *Economics of Innovation and New Technologies* **15**, 391–413. doi:10.1080/10438590500512950

NEFFKE F. M. H., HENNING M., BOSCHMA R. A., LUNDQUIST K. J. and OLANDER L. O. (2011) The dynamics of agglomeration externalities along the life cycle of industries, *Regional Studies* **45**, 49–65. doi:10.1080/00343401003596307

O'DONOGHUE D. and GLEAVE B. (2004) A note on methods for measuring industrial agglomeration, *Regional Studies* **38**, 419–427. doi:10.1080/03434002000213932

OTA M. and FUJITA M. (1993) Communication technologies and spatial organization of multi-unit firms in metropolitan areas, *Regional Science and Urban Economics* **23**, 695–729. doi:10.1016/0166-0462(93)90018-A

PORTER M. E. (1990) *The Competitive Advantage of Nations*. Free Press, New York, NY.

PUGA D. (2010) The magnitude and causes of agglomeration economies, *Journal of Regional Science* **50**, 203–219. doi:10.1111/j.1467-9787.2009.00657.x

RAMA R. and VON TUNZELMANN N. (2008) Empirical studies of innovation in the food and drink industry, in RAMA R. (Ed.) *Handbook of Innovation in the Food and Drink Industry*, pp. 13–49. CRC Press, Boca Raton, FL.

ROMER P. (1986) Increasing returns and long-run growth, *Journal of Political Economy* **94**, 1002–1037. doi:10.1086/261420

ROSENTHAL S. and STRANGE W. (2003) Geography, industrial organization, and agglomeration, *Review of Economics and Statistics* **85**, 377–393. doi:10.1162/003465303765299882

VAN DER PANNE G. and VAN BEERS C. (2006) On the Marshall–Jacobs controversy: it takes two to tango, *Industrial and Corporate Change* **15**, 877–890. doi:10.1093/icc/dtl021

VAN DER PANNE G. (2004) Agglomeration externalities: Marshall versus Jacobs, *Journal of Evolutionary Economics* **14**, 593–604. doi:10.1007/s00191-004-0232-x

VON TUNZELMANN N. and ACHA V. (2005) Innovation in 'low-tech' Industries, in FAGERBERG J., MOWERY D. and NELSON R. R. (Eds) *The Oxford Handbook of Innovation*, pp. 407–433. Oxford University Press, Oxford.

WOOD A. and ROBERTS S. (2011) *Economic Geography. Places, Networks and Flows*. Routledge, London.

WOOLDRIDGE J. M. (2002) *Econometrics Analysis of Cross Section and Panel Data*, MIT Press, Cambridge, MA.

Regional Knowledge Flows and Innovation Policy: A Dynamic Representation

UGO FRATESI
Politecnico di Milano, DABC, Piazza Leonardo da Vinci, Milan, Italy

FRATESI U. Regional knowledge flows and innovation policy: a dynamic representation, *Regional Studies*. This paper presents a tool to study *ex-ante* the effects of innovation policy on regional growth and income. It uses a dynamic evolutionary simulation approach and presents a model able to represent the flows of knowledge within and between regions. The model is unique but can be customized to represent different regional innovation modes. The model is calibrated with data of the average European Union NUTS-2 region, and is used to show the different impacts of various policy options, and the different impacts of the same policies in different regions, providing evidence in favour of regionally tailored, place-based approaches. Calibrating the model, through fieldwork, on an actual region will eventually allow *ex-ante* estimations of actual policy impacts to be produced.

FRATESI U. 区域知识流动和创新政策：一个动态的再现，*区域研究*。本文呈现一个事前研究创新政策对区域成长和所得的影响之工具。本文运用动态演化模拟方法，并呈现一个能够再现区域中和区域间的知识流动的模型。该模型是特殊的，但亦能客製化以再现不同的区域创新模式。该模型以欧盟一般NUTS-2层级的区域数据测定之，并用来展现各种政策选项的不同影响，以及同一政策在不同区域的不同影响，提供了支持为特定区域量身定做、根据地方的方法之证据。透过田野工作，根据一个真实的区域测定该模型，最终得以生产实际政策影响的事前评估。

FRATESI U. Les flux de connaissances et la politique d'innovation régionaux: une représentation dynamique, *Regional Studies*. Cet article présente un outil qui permet d'étudier ex-ante les effets de la politique d'innovation sur la croissance et le revenu régionaux. On se sert d'une simulation évolutive dynamique et présente un modèle capable de représenter les flux de connaissances à la fois intra et interrégionaux. Le modèle est unique mais on peut l'adapter afin de représenter divers modes d'innovation régionale. Le modèle est calibré avec des données sur la région type NUTS-2 de l'Union européenne et sert à montrer les différents impacts des diverses options politiques, et les différents impacts des mêmes politiques menées dans des différentes régions, ce qui a fourni des résultats en faveur des approches régionales faites sur mesure et adaptées au milieu. À partir du travail sur le terrain, calibrer le modèle sur la base d'une région réelle permettra au fur et à mesure de produire des estimations ex ante des impacts politiques réels.

FRATESI U. Regionale Wissensströme und Innovationspolitik: eine dynamische Darstellung, *Regional Studies*. In diesem Beitrag wird ein Instrument zur Vorabschätzung der Auswirkungen der Innovationspolitik auf das regionale Wachstum und Einkommen vorgestellt. Hierfür kommt ein dynamischer evolutionärer Simulationsansatz zum Einsatz, und es wird ein Modell zur Darstellung der Wissensströme innerhalb einer Region sowie zwischen Regionen vorgestellt. Dieses Modell ist spezifisch, lässt sich aber zur Darstellung verschiedener regionaler Innovationsweisen anpassen. Das Modell wird anhand von Daten der durchschnittlichen NUTS-2-Region in der Europäischen Union kalibriert und dient zur Verdeutlichung der unterschiedlichen Auswirkungen verschiedener politischer Optionen sowie der unterschiedlichen Auswirkungen derselben Politiken in verschiedenen Regionen, was Belege zugunsten von regional zugeschnittenen, ortsbasierten Ansätzen liefert. Eine Kalibration des Modells mithilfe von Feldarbeit in einer tatsächlichen Region ermöglicht langfristig eine Vorabschätzung der tatsächlichen politischen Auswirkungen.

FRATESI U. Flujos de conocimiento regional y política de innovación: una representación dinámica, *Regional Studies*. En este artículo se presenta una herramienta para analizar de forma *ex-ante* los efectos de la política de innovación en el crecimiento y los ingresos regionales. Se utiliza un enfoque de simulación dinámico evolutivo y se presenta un modelo capaz de representar los flujos de conocimiento entre las regiones y dentro de ellas. Este modelo es único pero puede adaptarse para representar los diferentes modos de innovación regional. El modelo se calibra a partir de datos sobre la región media NUTS-2 de la Unión Europea, y se utiliza para mostrar los distintos efectos de las diferentes opciones políticas, así como los distintos efectos de las mismas políticas en regiones diferentes aportando pruebas a favor de enfoques regionalmente adaptados en lugares determinados. Calibrar el modelo mediante un trabajo de campo en una región determinada permitirá a largo plazo hacer estimaciones *ex-ante* de los efectos políticos actuales.

INTRODUCTION AND AIMS

A considerable degree of diversity in regional innovation performances is empirically detected in studies such as the European Regional Innovation Scoreboard exercise (HOLLANDERS *et al.*, 2009), which also showed that regions have different strengths and weaknesses. Moreover, the innovation scores of the European regions examined in the exercise appear to be quite stable compared with the previous exercises of 2004 and 2006, signalling that regions are more and less innovative in nature. For instance, USAI (2011) showed that inventive performance at the Organisation for Economic Co-operation and Development (OECD) level is concentrated in regions characterized by human capital and research and development (R&D) expenditure. Indeed, regions appear in many cases to belong to different regional innovation models (CAPELLO and LENZI, 2015)

Empirically, it appears that the same innovation-related factors are not identically relevant for all regions. Moreover, the innovation-growth relationship is not equally important for all regions. STERLACCHINI (2008) showed that while highly educated people were beneficial to the gross domestic product (GDP) growth of all European regions, R&D expenditure was beneficial only for those above a certain threshold of GDP per capita. Additionally, the relationship between regional human capital and R&D and regional growth was evident for northern European countries but not for southern ones.

These theoretical findings are not easily reconciled with policy efforts, which are too often similar rather than differentiated between regions with different types of innovation systems (OECD, 2011). For instance, after the failure to achieve the Lisbon Strategy objectives, the new Europe 2020 strategy targets involves 'smart, sustainable, inclusive growth' and a target of 3% for the share of R&D expenditure on GDP. Within this strategy the flagship initiative 'Innovation Union' includes, among its objectives, 'Spreading the benefits of innovation across the Union', and states: 'The Innovation Union must involve all regions. [...] Europe must avoid an 'innovation divide' between the strongest innovating regions and the others' (EUROPEAN COMMISSION, 2010).

To implement this objective with a call for one-size-fits-all innovation policies would lead to lower performance than could be achieved with smarter, place-based, policies (BARCA *et al.*, 2012). This is particularly the case because the literature shows that innovation needs prerequisites (RODRÍGUEZ-POSE, 1999, 2001; CRESCENZI and RODRÍGUEZ-POSE, 2011); even where regions are apparently similar, new knowledge can only be built upon previously existing knowledge, so that innovation is a cumulative process localized at sectoral and territorial level (FAGERBERG, 2005).

This paper builds a tool to measure *ex-ante* the impact of various types of innovation policies at regional level on growth and income through their influence on intra- and inter-regional knowledge flows. A model is developed and calibrated on a generic European region in order to show the impact of different policies. The model, if calibrated on individual regions through extensive fieldwork, could also be used as a normative tool to assess *ex-ante* policies at the individual regional level. The paper will also show that different types of innovation policies have different impacts in time and depend on regional contexts, therefore supporting the need of region-specific policies.

The model presented in this paper is built using a dynamic evolutionary simulation approach first developed by FRATESI (2010), based on a system dynamics methodology. In this approach, 'soft' modelling is adopted where it is not possible to solve the model analytically and the possibility of optimizing the behaviour of agents is limited. However, in this approach, the simplifications required by traditional 'hard' modelling techniques are not needed, and it is possible to take into consideration concepts that are normally expressed in word terms, for example by heterodox theories. Heterodox deductions and theories can also be logically tested if formalized in this way, and when dealing with innovation at territorial level the advantages clearly overcome the limitations. In particular, it is helpful to represent loops and feedbacks that are theoretically evident but hard to represent with traditional techniques.

The simulation model built in this paper can represent the flows of knowledge within regions. The model is unique but designed to represent different regional innovation models – some flows can be switched off either exogenously or because some specific typologies of agents (such as researchers, inventors, educated people, creative ones, innovative firms) are too scarce in the region at the start of the simulation. Hence, it is possible to simulate the different mechanisms by which different types of regions can innovate and produce goods, and thus develop an understanding of their different roles in a knowledge economy.

The paper is organized as follows. The second section outlines the main traits of the methodology. The third section describes the model and the theory behind it. The fourth section shows how the model is calibrated to an 'average' European Union (EU) NUTS-2 region, using real data as far as possible. The fifth section presents some example simulations. The sixth section concludes and also presents the scope for further research.

THEORY AND METHODOLOGY

Most of the literature on innovation at the regional level is not formalized and follows three approaches:

- Conceptual studies, which are based on the capability of the researcher to identify, describe and, in some cases, classify the relevant concepts.
- Case studies, which are very interesting in the sense that they can go much deeper into the analysis of specific regional innovation systems, but are at the same time obviously limited in their ability to abstract the general from the particular.
- Empirical–econometric papers, in which hypotheses are made and tested through the use of statistical proxies for the concepts, quite often testing reduced forms.

This paper follows a different approach: it is formalized but uses a 'soft' modelling approach, developed below.

Five scales are needed to define the approach used in this paper. The first three come from BOSCHMA and FRENKEN (2006), who identify three dichotomies that are needed to classify a research approach in economic geography: (1) the use of mathematical formalization; (2) the assumption of the rationality of agents, which can involve full rationality or bounded rationality; and (3) the dynamic characteristics of the description, which can be static, based on comparative equilibria or, at the other end of the spectrum, fully dynamic, i.e., also investigating disequilibria trajectories.

The fourth dimension comes from CAPELLO (2007), according to which the conceptualization of space can be abstract and stylized (suitable for formalizations), or relational, when space is conceived as the support where many different sets of relations between economic agents can exist (in this latter case formalizing is more difficult). Finally, the fifth dimension concerns the scale of the arguments, which can range from macro to micro, with a so-called meso-approach between the two in which neither aggregates nor individuals are considered, but the focus is on institutions, groups of agents and their interactions (appropriate for studying, for example, routines, sectors and common practices).

With these dimensions in mind, the present study will use an approach based on evolutionary economic geography as defined by BOSCHMA and FRENKEN (2006): it is formalized, has bounded rationality and is essentially dynamic. The conceptualization of space aims at being relational; however, it is not possible to achieve this because of the need for formalization and the scale of arguments will therefore be essentially meso-economic.

A dynamic modelling methodology features the precious characteristic of interrelation: each economic sector can be linked with all other relevant ones and the representation of the production and diffusion of knowledge may include all the agents considered to be involved in these processes at the territorial level.

Many ways of dealing with simulation models are outlined in the literature, although more often in engineering or environmental studies than economics. Here, the chosen approach is a system dynamics one. The choice of system dynamics above other simulative approaches is due to the ease of achieving the aims of this work, especially because the approach allows two different phases to be separated out in the modelling process (WOLSTENHOLME, 1990):

- The design of the structure of the model according to the theoretical relationships which are considered most important by the modeller.
- The formalization of the equations according to the relationships that are specified in the first phase.

Separating out these two phases is very useful when dealing with very complex systems such as regional innovation systems: this is why, for example, the description of the relationships and the theory behind them is given in the third section, while the calibration and simulations are in the fourth and fifth sections of this paper.

Moreover, the explicit inclusion of feedbacks are characteristic of system dynamics, which is helpful when modelling regional systems: increases of production and employment in the region, for example, will lead to increased wages and production costs, which are detrimental to further increases of production and employment. Neglecting these feedbacks would lead to a model in which improving a node of the model means everything improves by definition, which is hardly compatible with reality.

Harder modelling techniques, such as agent-based models or computable general equilibrium models, could also be used for policy assessment. In comparison with these, the system dynamics approach is not optimizing because it is not based on individual agents taking rational choices. The advantage, however, is the fact that the time dynamics of flows can be represented and modelled clearly and consistently, and the targets and direct and indirect impacts of policies represented more effectively, so that the policy levers are explicit.

System dynamics, like other aggregate approaches, is exposed to the Lucas critique, whereby the behaviour of agents can be influenced by policies; at the same time, there is ample consensus in the literature that regional agents are not fully rational and optimizing, and since policies tested with a model like the one used in this paper have to be realistic, it is acceptable to think that realistic policies, being constrained in their size, do not significantly affect the behaviour of individual agents. It would be different if a policy were to use huge resources, so that it could radically change the regional economy; in that case, with a different local system, the behaviour of agents would certainly be affected. Unfortunately, however, the resources available for policies are never very large; for example, the EU budget hovers around 1% of EU GDP, and only about one-third is used for Cohesion Policy.

In the system dynamics approach the modelling process starts with the identification of stock variables, those persistent in time. These variables give inertia to the whole system because they maintain their value after each run of the simulation, unless affected by inflows or outflows (e.g., the population is affected by births and deaths, or patents are affected by new patent applications and by the expiration of existing ones). A model basically becomes a system of difference equations, where stocks evolve in time as follows:

$$\begin{aligned} \text{stock}_{t+1} &= \text{stock}_t + \text{flow}_t, \\ \text{where flow}_t &= \beta\,\text{stock}_t + \gamma_t \end{aligned} \tag{1}$$

In Fig. 2, stocks are represented with boxes and flows with thick arrows with a sort of hydraulic tap. The simpler relations, in which the value of a variable at time t is used to calculate the value of another variable at the same time t, are represented by thinner arrows. It is also possible to introduce variables that are neither time persistent nor directly affecting the stocks, if the modeller thinks they are important enough to be autonomous in the model rather than just values within the various equations.

The system dynamics methodology makes it compulsory to distinguish *ex-ante* which variables are persistent (stocks), which variables affect them (flows) and which variables are determined at each simulation run. This forces the modeller to be theoretically consistent; in the empirical literature, however, searching for regressors is often less rigorous, and different papers may use endowment and variation. There must therefore not be confusion between patenting versus stock of patents, R&D versus knowledge, infrastructure versus infrastructure investment, educated workforce versus attraction of graduates, and so on.

One important point to consider when analysing a dynamic system is the distinction between processes that change at different speeds. As first emphasized by the synergetics literature (HAKEN, 1983, 1985), and applied with success in models of transport, urban and regional evolution (ANDERSSON, 1986; BATTEN and JOHANSSON, 1987; JOHANSSON, 1993; JOHANSSON *et al.*, 2001), not all changes take place at the same pace and some slow-moving processes can act as an arena in which faster moving processes play their game. At the same time, perturbations in the fast-moving processes can lead to catastrophic changes in the slow-moving arena, if they are close to the bifurcation points. BARKLEY ROSSER (2011) distinguishes between three groups of changing processes in urban evolution: slow (e.g., industrial, residential and transport construction), medium (e.g., economic, demographic and technological change) and fast (e.g., labour, residential and daily mobility).

The model developed in the following sections is devoted to simulate policies to analyse short-to-medium-term policies, which act on regional competitiveness through an impact on the knowledge flows. Slow changing ('arena') processes, such as institutions, are normally inserted through parameters, i.e., are not directly modelled as endogenously changing, since their evolving time would be on a different scale with respect to the one of regional policy, but can be modified to simulate different regional milieus and hence the same policy in different regional contexts/arenas. Medium-speed processes are modelled through the persistence allowed by the stock and flow structure, so that, for example, an increase in the speed of patenting only slowly changes the stock of existing patenting, while fast processes are levers of the model which can be changed immediately.

As for all modelling methodologies, it is easy to succumb to the temptation to add complexity by including relationships that are realistic but not really necessary, making it harder to understand what stands behind the results obtained with the model. This is a common issue in science, and is described in Occam's razor, or its modern version, the 'verisimilitude trap' (GILBERT and DORAN, 1994).

The second step of the modelling process is to enter the actual equations, which is possible only according to the diagram previously designed, i.e., using the variables indicated by the arrows, with freedom to choose the best mathematical form for each equation.

In the modelling process, since time dynamics are very important, great attention has to be paid to the fact that different types of resources have different degrees of mobility between regions. For instance, consistent with the literature, different economic agents and people have to be assumed to be differently mobile (FRATESI, 2014b). Moreover, it is important to be aware of the fact that different processes work at different speeds. The model developed in this paper, for example, has a complex cohort-demographic mechanism for ageing and training, which takes into account the slow evolution of the processes concerning people; the policy simulations that are implemented also take account of these points, as it would be absurd to implement in one period a policy with an infrastructure increment of 20%. On the contrary, it is possible suddenly to increase the public research budget by 20%, even if it takes time as money goes into research, then patents and then products.

This dynamic methodology was introduced by FRATESI (2010), where its advantages are critically discussed with the help of a theory-based model. The present study is an ambitious further step insofar as it aims to draw policy conclusions for actual regions rather than theoretical ones. For this reason, a model is built, consistent with the theory, in which actual regional values can be entered rather than theoretical ones.

In this paper the model is built at the regional level, and calibrated on the actual values of an average EU NUTS-2 region. Another possibility would have been to calibrate it on a functional urban region. The latter option would have brought the advantage of being more consistent with the theory, having a self-contained labour market; NUTS-2 regions often contain more than one labour market, and in some cases share some local labour markets with neighbouring regions. The choice of NUTS-2, however, is justified by the fact that the model is to be used as a policy tool, and hence needs to be calibrated on a unit with policy relevance. For some European countries (e.g., Italy and Spain), NUTS-2 regions coincide with administrative units, while for other countries, where administrative units are different (e.g., Germany, where the *landers* are NUTS-1) NUTS-2 still has meaning as the unit of measurement in which eligibility for EU Structural Funds is calculated. Currently (and perhaps unfortunately), functional urban areas that exist in economic terms are very rarely endowed with administrative powers.

THE MODEL

General features

The model developed in this paper builds on a number of stylized facts taken from the very wide literature on innovation at regional level. In particular, inspiration is taken from studies that view innovation as an evolutionary process (NELSON and WINTER, 1982; LUNDVALL, 1992; NELSON, 1993; EDQUIST, 1997), studies on clusters and regional innovation systems (ASHEIM *et al.*, 2006; BRACZYK *et al.*, 1998; LUNDVALL, 1992), studies on the recently labelled 'Evolutionary Economic Geography' (BOSCHMA and LAMBOOY, 1999; BOSCHMA and FRENKEN, 2006; KOGLER, 2015), studies on knowledge spillover and other mechanisms for the interregional transfer of knowledge (AUDRETSCH and FELDMAN, 1996, 2004; BRESCHI and LISSONI, 2001, 2009), and on the role of socio-economic factors as a filter for innovation at regional level (CRESCENZI and RODRÍGUEZ-POSE, 2011).

The model is also clearly inspired by cumulative causation models (MYRDAL, 1957; KALDOR, 1970; DIXON and THIRLWALL, 1975) and adopts a territorial capital perspective in which a region is studied in its material and immaterial, private and public assets (CAMAGNI, 2009).

The model built for this paper represents the basic structure of a regional knowledge economy, and can be expanded to allow for further complexity or different aspects where necessary, such as for application to a specific region, or to unbundle some sectors from the others for the study of sectoral-specific policies.

As shown in Fig. 1, the model is built on two main elements:

- An internal knowledge-circulation dimension.
- An internal–external knowledge dimension.

The internal knowledge-circulation dimension is built on three main stocks: knowledge, patents and high-level products (lower-level standardized products are less important in the knowledge economy so they are modelled in lesser detail). Knowledge circulation tends to follow a 'refined linear model' (BALCONI *et al.*, 2010) in which traditional mechanisms are present but interactions – more recently introduced in the literature – are also taken into account.

The internal–external dimension is reflected by the activation of flows between the region (whose boundary is represented by the dotted circle in Fig. 1) and the external world. There are in fact a number of flows linking the region and the rest of the world, essentially concerning four groups of assets: knowledge, people,

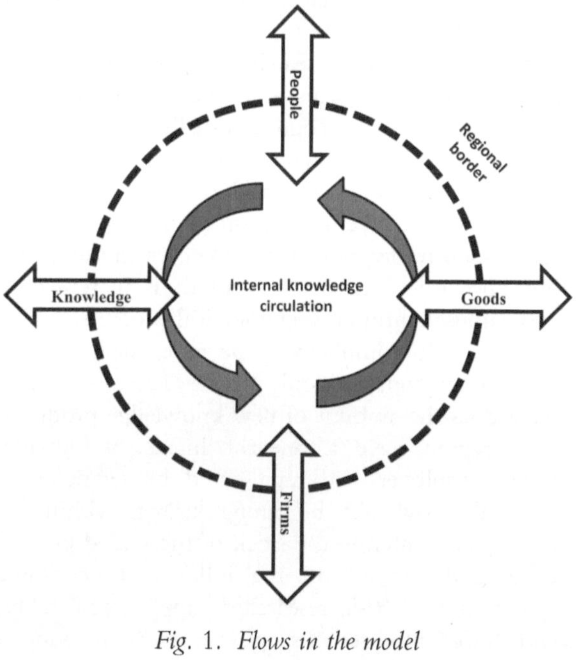

Fig. 1. Flows in the model

goods and firms. The three material flows can also carry knowledge, which is transferred with people, firms and goods. Especially relevant are the flows of people and firms, since these bring with them the possibility of the transfer of knowledge, consistent with the most recent literature on spillover, which has found the mechanisms underlying spillover in the mobility of personnel (e.g., MOEN, 2005). However, trade also brings possibilities for learning, since people and firms can learn from imported goods or the procedures of the firms exporting them.

From knowledge flows to patents to products

The model includes explicit and tacit knowledge. The first is modelled through three interacting stocks (see the lower left of Fig. 2; and see Fig. A1 in the supplemental data online):

- *Internal knowledge*, which is exclusive to the region and is available elsewhere. This is increased by basic research, which is performed by basic researchers but assumed to be facilitated by the presence of applied researchers in the region.[1] This is consistent with studies that see a role for business–university interactions in the construction of knowledge (LEYDESDORFF and ETKOWITZ, 1998); moreover, the public research budget is assumed to be relevant to the ability of basic researchers to create new knowledge, as research and laboratories need funding.
- *Shared knowledge*, which includes explicit knowledge not exclusive to the region but shared with other regions. Shared knowledge is incremented by outward spillover from the region, which makes knowledge that was created internally available also externally, and inward spillover (described below). The magnitude of outward spillover is dependent on the stock of internal knowledge, but also on the outward flows of three types of agents able to carry knowledge – basic researchers, applied researchers and entrepreneurs/firms – as well as openness to trade. Shared knowledge is usually relatively old since it has been invented somewhere and afterwards 'copied' elsewhere, so the stock of shared knowledge also has a dissipative mechanism of obsolescence.
- *External knowledge*, which is not known in the region but known elsewhere. Its growth is incremented exogenously through technological progress, which is assumed for simplicity to be independent of the modelled region, since only in a very few cases worldwide does the amount of new knowledge produced in a region have a relevant impact worldwide. Inward spillovers are represented by elements of external knowledge becoming known within the region, i.e., entering the stock of the shared knowledge, and depend on the inflows of economic agents such as basic researchers, applied researchers and firms/entrepreneurs; private research budgets are also assumed to be relevant, since they help reverse engineering, patent acquisition and other processes which support the acquisition of external knowledge. Finally, the more a region is open to trade, the more likely are inward spillovers.

With the three stocks of knowledge a synthetic indicator is built to measure the ability of the region to invent new things, which depends not only on how much knowledge the region possesses but also on its share of the knowledge possessed worldwide. The indicator is called the 'knowledge gap' and ranges between 1 (where all world knowledge is known in the region) and infinity (where the region knows nothing), and is defined as follows:

$$\text{knowledge gap} = \frac{\text{total knowledge}}{\text{regional knowledge}} = \frac{\text{external knowledge} + \text{internal knowledge} + \text{shared knowledge}}{\text{internal knowledge} + \text{shared knowledge}} \tag{2}$$

Still following a refined linear model, the knowledge gap is the basis of the regional ability to produce new patents, which in the model are also a proxy of market-usable regional knowledge. The model therefore applies a process in which knowledge is used to produce patents and patents are used to invent and introduce products (see the upper part of Fig. 2; and see Fig. A2 in the supplemental data online) involving two main groups of economic agents: applied researchers and entrepreneurs.

Creating new patents from regional knowledge requires applied researchers as well as the intervention of research funds (which in the simplification of the model are assumed to be coming exclusively from the private sector), and the presence of the right institutions, consistent with theories of the so-called 'social filter' (RODRÍGUEZ-POSE, 1999). It is a known fact that some regions possess institutions that facilitate knowledge flows within regions and better allow knowledge to translate into new products, while other regions have institutions that are innovation harmful (CRESCENZI and RODRÍGUEZ-POSE, 2011). These are the institutions that are most relevant in the context of this paper. Since patents expire and their usefulness decreases as technology progresses and new techniques replace older ones, they are modelled to have a mechanism of obsolescence.

Patents do not transform automatically into products that can be sold in markets, but need a mechanism involving uncertainty (which can be modelled stochastically[2]) and the intervention of specific agents, i.e., entrepreneurs. The action of entrepreneurs is made easier by the possibility of obtaining financial capital to fund their entrepreneurial initiatives. For traditional firms this means low real interest rates and, for the

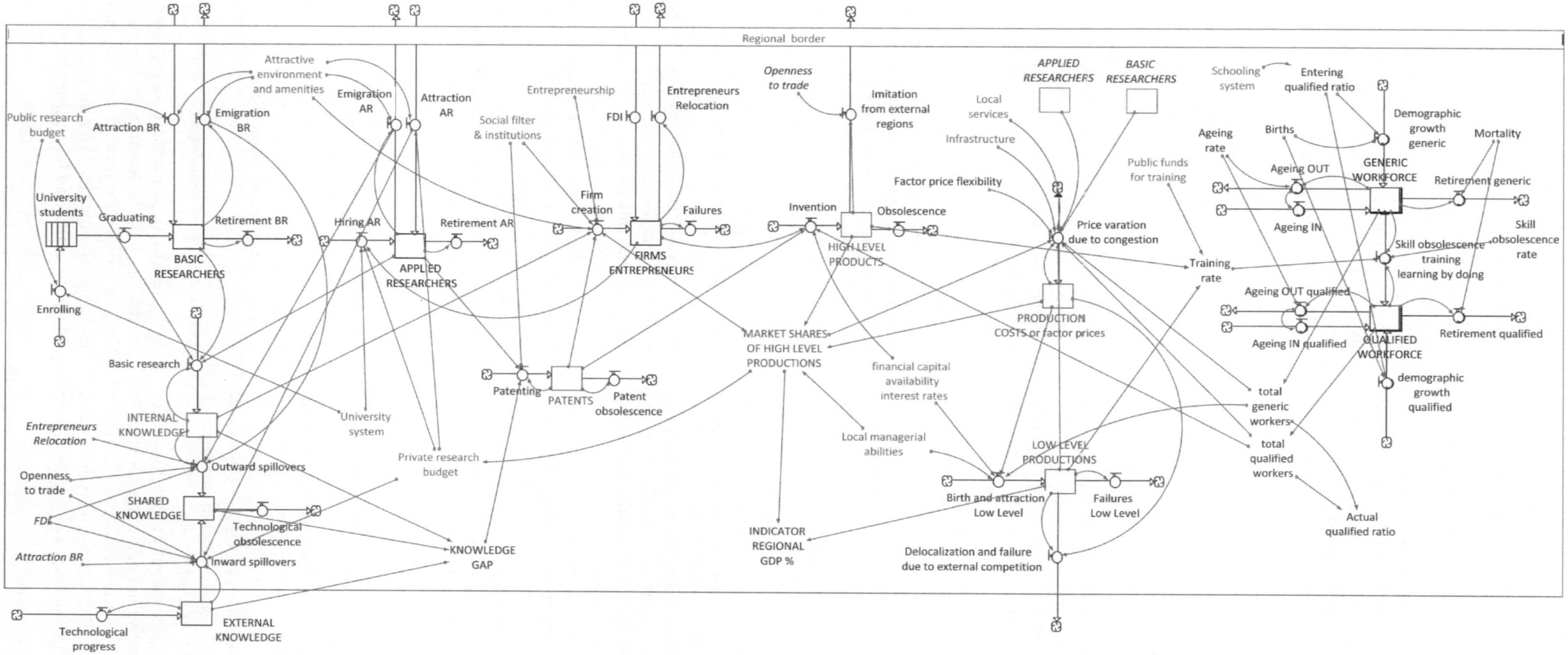

Fig. 2. The full diagram of the model (enlargements are in Appendix A in the supplemental data online)

most innovative entrepreneurial activities, the presence of venture capital.

'High-level products' are modelled as a stock since, even with shortening product life cycles, every product stays in the market for a while before being replaced by others. Since products are subject to obsolescence and external competition, their stock needs to be continuously replenished. If the region is more open to trade, it is easier for competitors to imitate products.

All processes of knowledge creation should be modelled at least stochastically, since all innovation involves uncertainty (NELSON and ROSENBERG, 1998). However, if the purpose is to use this typology of models for *ex-ante* policy assessment, the 'expected outcome' of policies is the most relevant variable, and the use of confidence intervals is rarely seen in this kind of exercise. For this reason, the model is calibrated deterministically and its results, presented below, are deterministic, although there is a fully functioning version of the model in which stochastic results of the process are included.

The economic agents involved in the knowledge and economic processes

As stated above, three sets of agents are explicitly involved in the knowledge economy: basic researchers, applied researchers and entrepreneurs (Fig. 2). All three are mobile between regions and can be attracted to or relocate/emigrate out of a region in search of better conditions. An attractive environment and the presence of amenities is a determinant of the inflows of all three. While all three sets of agents are mobile between regions, they are not assumed to be equally mobile, as their migration parameters are different – for example, the entrepreneurship literature shows that entrepreneurs mainly start firms in the region where they live – and they need networks to be effective in their activities (FISCHER and NIJKAMP, 2009).

Basic and applied researchers are also subject to 'retirement' processes in order to model the fact that these agents also have a working lifespan after which they either retire or lose their creative and innovative impetus. For entrepreneurs, age is not as important as the ability to be creative and innovative, as shown by the highly non-uniform distribution of the ages of entrepreneurs owning start-ups. The regional ability to educate and hire basic and applied researchers is affected by the level of schools and universities, their educator role still being highly relevant (VARGA, 2009), while the public research budget is also a factor in attracting basic researchers in search of funds.

As already mentioned, the mobility of specific agents is a clear channel of inbound and outbound knowledge flows, and, following recent empirical literature (BRESCHI and LISSONI, 2009), the model developed here accounts for the interregional mobility of knowledge through the mobility of specific groups of people. For example, attracting basic researchers facilitates the expansion of the stock of regional knowledge (Fig. 2). The model could in future be extended to become a multi-regional model by linking regions through flows.

The modelling of the tacit knowledge mechanisms (first introduced in the literature by POLANYI, 1969; and ROSENBERG, 1982) and learning by doing are captured in the model by the capabilities of workers. These can represent either a 'generic workforce', only possessing generic abilities, or a 'qualified workforce', with specific abilities that cannot be learnt at school (see the right of Fig. 2; see also Fig. A3 in the supplemental data online). The two stocks of workers start therefore from the same source (i.e., demographic growth), but the quality of schooling decides the share of new workers who are generic and need training and/or learning by doing to become qualified – probably the larger part.

Both generic and qualified workers have a full demographic mechanism of ageing, with people in 40 separate cohorts (assuming 40 years as the maximum working life of people) and a retirement mechanism that starts acting with increasing rates with the increase of age. In Fig. 2 this is represented, synthetically, with overlapped squares, meaning that these are not simple stocks but matrixes of stocks (in this case 40 stocks, one per age group).

More relevant to the knowledge economy, people in the model not only age in each period but also can acquire skills and expertise, or become unskilled if their expertise becomes obsolete – skills are also subject to obsolescence and need to be renewed and replaced, and this normally happens at rates increasing with age. This is modelled with a bi-directional mechanism of obsolescence and training/learning by doing, linking the two stocks of qualified and generic workforce, in which various factors influence the percentage of people of each age group that gains skills or becomes unskilled. These factors are assumed to be the public funds for training and the presence in the region of – especially high-level – production.

This complex demographic mechanism is important because if people are trained when they are close to retirement, they will be able to use their increased expertise only until they exit the labour market. In this way, the cohort-demographic mechanism can also consider the presence in the region of a younger or an older population and labour force.

Congestion, factor costs and regional wealth

With respect to congestion, factor costs and regional wealth, a dynamic system approach proves useful. In fact, all the flows represented in the previous sections can trigger positive loops for the region, but it is impossible in the real world for a region to become too big in economic terms with respect to its population and size, and it is also difficult to become drastically richer than other regions, owing to the pressure on factor prices and the insurgence of congestion diseconomies after a certain threshold. Without compensating mechanisms, the cumulativeness of knowledge, with ARTHUR's (1994) increasing returns to scale, would lead to indefinite growth inside the region as a consequence of increasing returns in the innovation process. The dynamic methodology allows one to model the negative feedbacks from success to increased costs, so that the attractiveness of the region for further economic activities is counterbalanced by larger production costs.

The model includes a 'congestion' mechanism (see the centre of Fig. 2; see also Fig. A4 in the supplemental data online), where congestion is due to the presence of people and firms in the region, meaning slightly greater weighting for those people and firms which are likely to be richer and hence consume more resources. Congestion is the determinant of local 'production costs', where factor prices increase with the pressure on them. Production costs are represented with a stock for a technical reason: this allows a delay between the increase or decrease of congestion and the increase or decrease of factor prices. Prices are sticky, for example, because wages are set for the length of contracts, and hence it would be unrealistic to assume that they vary instantaneously. Prices hence adjust to congestion as follows:

$$\text{prices}_{t+1} = \text{prices}_t + adj * (\text{Prices with congestion}_t - \text{prices}_t) \quad (3)$$

where *adj* (between 0 and 1) is the speed of adjustment and actual prices adjust to prices with congestion in time. By varying the value of *adj*, it is possible to model different regional situations in which prices are more or less flexible. This price stickiness is important in the simulations because it is an important determinant of the speed of feedbacks in the economy, and hence of the time delay of policies.

Production costs and the number of high-level products developed by the regional knowledge mechanisms are the two main determinants of the regional market shares of high-level products. Firms will decide to produce high-level goods and services in the region unless high production costs convince them that it is convenient to shift production elsewhere. Local managerial abilities can influence these location choices. The presence of firms producing high-level products in the region positively influences the amount of private resources devoted to regional innovation, thus feeding the knowledge production mechanisms.

No region produces only high-level products; traditional low-level goods and services are also produced. Being marginal to the knowledge economy, these are modelled in less detail (see the lower part of Fig. 2); as they are non-innovative, the competitiveness factors are also different and clearly more closely linked to traditional location factors such as production costs, congestion, the presence of a labour force, and the availability of cheap financial capital for investments and of local managerial competences.

Both high- and low-level production (though the former does so to a larger extent) contribute to the level of income per capita in a region – which is the final indicator of the model, although it cannot capture the complexity of economic development. It should be clear that the model is thus not a generative growth model but a 'competitive' one, in RICHARDSON's (1978) terminology. The absolute growth rate is not determined in the region, and nor is growth at the world scale; however, the region can place itself lower or higher in the ranking of regions depending on how well knowledge-creation mechanisms work.

As a model whose aim is to simulate the impact of policies, this is not a shortcoming, since a policy-maker needs to evaluate a policy against a counterfactual of 'no policy', or to test one policy against another. The absolute growth rate of an economy does not depend on the region itself, but the region can be more or less sensitive to the movements of the national and global economy. For example, almost all regions lost jobs in the current economic crisis, but those with better knowledge mechanisms and better policies have lost the fewest.

CALIBRATION OF THE MODEL

Three options are possible to calibrate the model before using it for policy simulations.[3] One option is to calibrate it on theoretical values, which works better if all values are standardized, because in this way the values are immediately comparable. This is the option taken in the earliest attempts with this methodology (FRATESI, 2010). This method is relatively easy and can lead to important theoretical results, but these will also be most abstract and difficult to interpret from a policy point of view.

Another option is to calibrate the model on the actual values of a specific region. However, this would be very hard to implement since the exact values of the parameters in a region, especially sensitivity parameters, are impossible to obtain without specific and intensive fieldwork. In addition to the practical difficulties, calibration with data on one region would only produce results valid for that specific region, which would be difficult to generalize.

In this paper a third, intermediate, option is used, which is better suited to analyse general policy issues and to show the importance of place-based policies. The model is calibrated, whenever the data allow, on a generic European region, using data mainly from European statistical sources, such as the Regional Innovation Scoreboard Exercise (HOLLANDERS *et al.*,

2009).[4] The values for this generic region, as mentioned above, are assumed to be the average of European NUTS-2 ones. The calibration on a generic region will form the benchmark on which other types of regions can be represented, starting from the generic region and changing the values of the relevant nodes (stocks and parameters, especially those determining the arena, i.e., the local economic milieu) when needed, in order to analyse the impact of policies in different types of region.

The choice of representing regional GDP with a per cent of the EU average, rather than an absolute value, comes from the focus of the model on supply-side aspects. While demand-side aspects are indeed important, this method allows one to neglect macroeconomic cycle issues, such as the economic crisis, which affect all regions, though in different ways. Region-specific demand issues, for example due to sectoral specialization, can be included in the model at a later stage, when policy-makers need to see the impact of sector-specific policies.

Those parameters not available from statistical data can be estimated by (expensive) fieldwork once the model is applied to a specific region. In this pilot exercise, they are set to plausible values, in particular assuming that the elasticity of one variable to another, as one among several influences, is smaller than 1, and that the elasticity with respect to more important variables is larger than in relation to less important variables.

The model is non-linear, but is calibrated in a stable equilibrium, and all simulations take place in the region of the same equilibrium, which is justified by the fact that policies are normally incremental.

Thanks in particular to the time delays of the model (due to price rigidity, the time needed to build stocks of items such as patents and the inertia in the demographic mechanisms), it will be possible to observe the dynamics of policies, since any policy needs time delays:

- to impact on the direct target; and
- to affect the whole regional economy.

Being systemic, this approach can also take account of indirect and recursive, second- and third-order effects, as already mentioned. In this way it is also possible to observe which policies are more effective in different types of region and which can produce more immediate impacts.

POLICY SIMULATIONS AND RESULTS

The aim of this section is to present the potential of the instrument with some illustrative policy simulations and some important issues related to the policies, in particular concerning the different impacts of conjunctural versus structural policies, with the former defined as those that temporarily alter the status of the regional system without altering its long-term equilibrium, while the latter are defined as those policies aimed at altering this long-term equilibrium (i.e., for what possible, the 'arena'). Attention is also paid, as favoured by a dynamic approach, to the time impacts of policies, which are generally inadequately considered by policy-makers, probably because of the scarcity of instruments to carry out relevant analysis.

This type of model can produce an almost unlimited number of policy simulations so a few examples are selected here in order to show the general conclusions on the utility of this type of model. In particular they demonstrate the utility of analysing a policy's quantitative impact before its implementation, and the time delay of this impact at the same time. It is less meaningful at this stage to compare two policies based on the absolute size of their impact, because it is easy to achieve a higher impact by introducing a larger change in the policy variable. This would only be relevant at a later stage in which a model such as this could be applied to a real region rather than a standardized one.

In addition to the above, other simulations show the different impact of policies in different contexts. Applying policies to different regional contexts can also obtain very differentiated results, also from an aggregate point of view (FRATESI, 2008), an outcome recognized by supporters of place-based policies (BARCA *et al.*, 2012) but generally not modelled due to practical difficulties.

One key question for any policy simulation is the cost of policies, since public finances are increasingly being constrained, especially as a result of the economic crisis and, in Europe, the Stability and Growth Pact. However, regions normally have no or very limited taxation power, and most of their resources come from higher bodies such as the nation-state or the EU. For this reason, rather than including a mechanism of local taxation (which would be possible, making the model more complex and less realistic), this model instead concentrates on the impact of policies, assuming that a region has a budget to use. This is, for example, how the Structural Funds work in Europe: the amount is decided at the beginning of the seven-year programming period and the money is then used according to programming documents.

The first simulation is the comparison of two conjunctural policies to show not only that two policies can have impacts of different magnitude but also that even when the absolute magnitude of the impact is the same they may have different time impacts, which makes it impossible to compare them by just looking at a single indicator without observing their time effects (Fig. 3(a)).

In this case there are two different positive shocks inserted in the model, one that temporarily decreases the production costs (such as temporary de-taxation of work) and one that temporarily increases basic research in the region for a limited amount of time (such as temporary research incentives). Since basic research needs time before patents are developed and products enter the market, the first policy has a more immediate impact, which, however, fades with the end of the incentive.

The second simulation concerns the comparison of two structural policies, which produce long-term changes, although again with different time effects (Fig. 3(b)). In particular, in this case the two policies are a permanent reduction of the cost of capital in the region (e.g., with a better implementation of the rule of law) compared with a policy that makes educational institutions produce qualified workers at a higher rate. The latter needs time to affect the labour force, starting from the first cohort of workers, but the former has a more immediate impact. Moreover, due to increased wages, the latter also has an 'overshooting' effect.

The third simulation compares a structural and a conjunctural policy (Fig. 3(c)). The structural policy is in this case a permanent increase in the level of urban amenities, which, in time, would favour the attraction of creative and knowledgeable people, and therefore also inward spillovers and regional innovation. The conjunctural policy is a temporary reduction of the cost of capital (e.g., driven by de-taxing capital invested in a

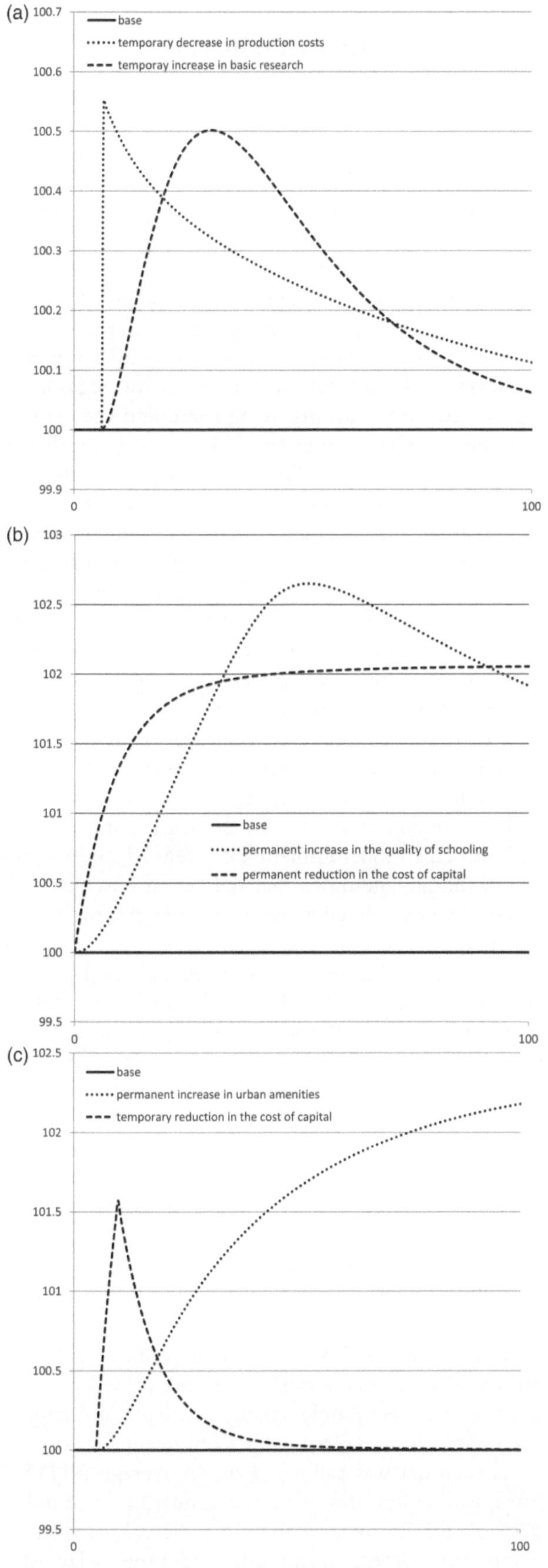

Fig. 3. Examples of policy simulations with the model

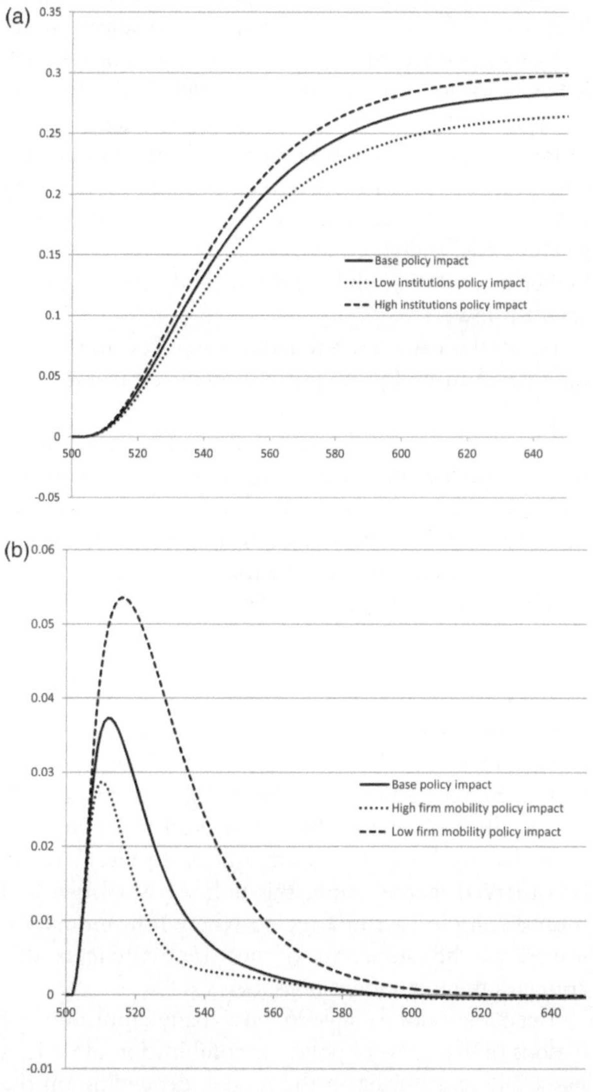

Fig. 4. Impact of the same policy in different regional situations

region). Although the structural policy is in this case to be preferred in the long run, it is clear that it cannot pull a region forward in the short run, and for this reason the best option to ensure an immediate and lasting impact would be a mix of the two policies.

Comparing policies in the same regional situation is just one possibility allowed by the model. In fact, regional scholars are increasingly in favour of place-based policies, being convinced that policies need to be targeted at the specific regions in which they are deployed in order to be effective (BARCA *et al.*, 2012). For this reason, Fig. 4 models the impact of the same policy in different regions.

The first experiment deals with the impact of a 10% increase of public research budget (say, new funds for research projects) on the indicator 'GDP%' (Fig. 4(a)). This is a policy that needs a long time before being effective, since it must move from knowledge to patents and then products. What is really interesting to observe, however, is its impact in different contexts. The base simulation is the policy applied to the standard calibration region, the one at 100% of the EU average.

The other two simulations apply to a region with a high level of institutions, which is innovation prone, and to a region with a low level of institutions, which is less innovation prone. In the model, institutions act as a social filter, which allow basic knowledge to be more easily applied and become patents, and prospective entrepreneurs to start-up businesses more easily using new ideas and new patents. In the first case, *ceteris paribus*, the regional equilibrium with no policy is higher than 100%, while in the second case the equilibrium is lower.

Given this paper's interested in the policy impact, the differential made by the policy is depicted in Fig. 4(a), rather than the absolute value of the indicator. As can be easily observed, the same increase of the public research budget produces a larger impact in a region with more innovation- and entrepreneurship-friendly institutions, while the impact of the policy is weaker when such institutions are themselves weak.

The second experiment concerns attracting external entrepreneurs to invest in the region (Fig. 4(b)). This could be done, for example, with lump-sum incentives. In this case, the policy is a temporary one, lasting only five periods. The policy is now applied to three regional cases, which are all at the same level of starting GDP (100%) but are different in the mobility of firms, with one having higher firm birth and death rates, and the other having lower firm birth and death rates. As can be observed in the graph, this policy has a larger and more lasting impact in a region where firm mobility is lower, as the attraction of new entrepreneurs discourages the incumbent ones less.

Even if it is not possible to show many simulations for reasons of space, every policy or combination of policies has a different impact in the model, depending on the starting regional situation. This makes the instrument fully place-based, i.e., in line with recent theories which affirm that policies have to be tailored to each region.

CONCLUSIONS

In an era in which it is often suggested that regional policy should become increasingly place based (BARCA, 2009), *ex-ante* policy assessment needs new tools that can deal with the diversity of regional situations and, hence, can simulate the impact of policies in different regional contexts.

It is not just a matter of comparing different policies but also of considering whether the same policy has the same impact if implemented in different regions. The receptiveness to policies in fact varies significantly with the level of the institutions present in the region, the type of economic agents living there and the type of economic activities present. To use a term that is becoming increasingly diffuse, one could say that the receptiveness to policies is linked to the level of territorial capital, which is changing but only through very slow processes. Also important are the composition of territorial capital and the typology of the policies being implemented. Territorial capital has many facets, and these can act as catalysts for different regional policies, with each territorial capital asset being able to expand the impact of some types of policies.

This paper has proposed one tool to assess *ex-ante* place-based policies, a simulation evolutionary economic geography model showing the flows of knowledge within a region and between the region and external regions. This model implements a refined linear innovation model, adding to that the recent idea that spillovers do not just take place by pure proximity but generally through the movements of specific groups of people, such as entrepreneurs or researchers (BRESCHI and LISSONI, 2009). A simulation methodology is useful because it allows a representation of the complexity of theoretical arguments without oversimplification, and at the same time allows the sound testing of theoretical hypotheses. Moreover, a simulation approach allows modelling of interrelation and recursiveness.

The model developed in this paper is built upon three stocks of knowledge – regional internal knowledge, shared knowledge and external knowledge – which allow a representation of interregional knowledge flows. Within a region there is another process in which knowledge is used to generate patents, which, in turn, can be used to invent new products. All these mechanisms are made possible by the presence of the most relevant economic agents, i.e., basic and applied researchers, firms/entrepreneurs and qualified workers.

The model was calibrated on an average NUTS-2 European region, this being the standard unit of policy decision for European cohesion policy. This allows a comparison of the magnitude and time delay of a number of possible policy interventions. It is important to understand the time dynamics of policies before implementing them, so that the impacts of both conjunctural and structural policies can be considered in different cases. By modelling different regional situations through different parameters of the model, the paper has also shown that, as advocated by the theorists of the place-based approach, the same policy can have very different impacts in different contexts.

The next step in the research is the adaptation and application of the model to a specific region. This requires calibration on actual data, and most likely extensive fieldwork carried out with a public administration, which makes this step challenging enough to be pursued as a separate stage. The result of this next step will be a useful *ex-ante* appraisal tool for policy-makers, who will be able to compare the impact of different policies in their territories. The framework is general enough to be extended to analyse sector-specific policies or to disaggregate regional institutions. The final aim is to facilitate better competitiveness and innovation policies more suited to the characteristics of the regions to which they are applied.

Disclosure statement – No potential conflict of interest was reported by the author.

Supplemental data – Supplemental data for this article can be accessed at http://10.1080/00343404.2015.1068930

NOTES

1. For simplicity, this paper considers 'basic researchers' to be primarily involved in the creation of knowledge, and 'applied researchers' in the patenting process, even if interactions between the two do exist and it is likely that both types of researchers produce both outputs, even if to different extents.
2. Stochasticity is handled in the model by making random rather than deterministic (with the same mean and a realistic variance) efforts in relation to processes such as patenting, but this is to move a little too far from this paper's focus on policies.
3. The choice of a region instead of a functional urban area was discussed in the second section.
4. A previous, intermediate, step of the research includes the detail of the actual sources and values used (FRATESI, 2014a).

REFERENCES

ANDERSSON Å. E. (1986) The four logistical revolutions, *Papers in Regional Science* **59**, 1–12. doi:10.1111/j.1435-5597.1986.tb00978.x

ARTHUR W. B. (1994) *Increasing Returns and Path Dependence in the Economy*. University of Michigan Press, Ann Arbor, MI.

ASHEIM B., COOKE P. and MARTIN R. (Eds) (2006) *Clusters and Regional Development: Critical Reflections and Explorations*. Routledge, London.

AUDRETSCH D. B. and FELDMAN M. P. (1996) R&D spillovers and the geography of innovation and production, *American Economic Review* **86**, 630–640.

AUDRETSCH D. B. and FELDMAN M. P. (2004) Knowledge spillovers and the geography of innovation, in HENDERSON J. and THISSE J. (Eds) *Handbook of Regional and Urban Economics*, pp. 2713–2739. North Holland, Amsterdam.

BALCONI M., BRUSONI S. and ORSENIGO L. (2010) In defence of the linear model: an essay, *Research Policy* **39**, 1–13. doi:10.1016/j.respol.2009.09.013

BARCA F. (2009) *An Agenda for a Reformed Cohesion Policy: A Place-Based Approach to Meeting European Union Challenges and Expectations*. Independent Report prepared at the request of Danuta Hübner, Commissioner for Regional and Urban Policy, European Commission.

BARCA F., MCCANN P. and RODRÍGUEZ-POSE A. (2012) The case for regional development intervention: place-based versus place-neutral approaches, *Journal of Regional Science* **52**, 134–152. doi:10.1111/j.1467-9787.2011.00756.x

BARKLEY ROSSER J. (2011) *Complex Evolutionary Dynamics in Urban–Regional and Ecologic–Economic Systems: From Catastrophe to Chaos and Beyond*. Springer, Berlin.

BATTEN D. and JOHANSSON B. (1987) Dynamics of metropolitan change, *Geographical Analysis* **19**, 189–199. doi:10.1111/j.1538-4632.1987.tb00124.x

BOSCHMA R. and FRENKEN K. (2006) Why is economic geography not an evolutionary science? Towards an evolutionary economic geography, *Journal of Economic Geography* **6**, 273–302. doi:10.1093/jeg/lbi022

BOSCHMA R. A. and LAMBOOY J. G. (1999) Evolutionary economics and economic geography, *Journal of Evolutionary Economics* **9**, 411–429. doi:10.1007/s001910050089

BRACZYK H. J., COOKE P. and HEIDENREICH M. (Eds) (1998) *Regional Innovation Systems*. UCL Press, London.

BRESCHI S. and LISSONI F. (2001) Localised knowledge spillovers vs. innovative milieux: knowledge 'tacitness' reconsidered, *Papers in Regional Science* **80**, 255–273. doi:10.1007/PL00013627

BRESCHI S. and LISSONI F. (2009) Mobility of skilled workers and co-invention networks: an anatomy of localized knowledge flows, *Journal of Economic Geography* **9**, 439–468. doi:10.1093/jeg/lbp008

CAMAGNI R. (2009) Territorial capital and regional development, in CAPELLO R. and NIJKAMP P. (Eds) *Handbook of Regional Growth and Development Theories*, pp. 118–132. Edward Elgar, Cheltenham.

CAPELLO R. (2007) *Regional Economics*. Routledge, London.

CAPELLO R. and LENZI C. (2015) Knowledge, innovation and productivity gains across European regions, *Regional Studies*, **49** (11). doi:10.1080/00343404.2014.917167

CRESCENZI A. and RODRÍGUEZ-POSE A. (2011) *Innovation and Regional Growth in the European Union*. Springer, Berlin.

Dixon R. and Thirlwall A. P. (1975) A model of regional growth rate differences on Kaldorian lines, *Oxford Economic Papers* **27**, 201–214.

Edquist C. (Ed.) (1997) *Systems of Innovation: Technologies, Institutions and Organizations.* Pinter, London.

European Commission (2010) *Europe 2020: A European Strategy for Smart, Sustainable and Inclusive Growth.* 3 March, COM(2010) 2020. European Commission, Brussels.

Fagerberg J. (2005) Innovation: a guide to the literature, in Fagerberg J., Mowery D. C. and Nelson R. R. (Eds) *The Oxford Handbook of Innovation*, pp. 1–26. Oxford University Press, Oxford.

Fischer M. and Nijkamp P. (2009) Entrepreneurship and regional development, in Capello R. and Nijkamp P. (Eds) *Handbook of Regional Growth and Development Theories*, pp. 182–198. Edward Elgar, Cheltenham.

Fratesi U. (2008) Regional policy from a supra-regional perspective, *Annals of Regional Science* **42**, 681–703. doi:10.1007/s00168-007-0167-x

Fratesi U. (2010) Regional innovation and competitiveness in a dynamic representation, *Journal of Evolutionary Economics* **20**, 515–552. doi:10.1007/s00191-009-0169-1

Fratesi U. (2014a) Simulating the impact of regional innovation policies with a dynamic model of regional knowledge flows, in Patrucco P. P. (Ed.) *The Economics of Knowledge Generation and Distribution: The Role of Interactions in the System Dynamics of Innovation and Growth*, pp. 252–271. Routledge, Abingdon.

Fratesi U. (2014b) Editorial: The mobility of high-skilled workers – causes and consequences, *Regional Studies* **48**, 1587–1591. doi:10.1080/00343404.2014.955689

Gilbert N. and Doran J. (Eds) (1994) *Simulating Societies.* UCL Press, London.

Haken H. (1983) *Synergetics, Nonequilibrium Phase Transitions and Social Measurement*, 3rd Edn. Springer, Berlin.

Haken H. (1985) Synergetics – an interdisciplinary approach to phenomena of self-organization, *Geoforum* **16**, 205–211. doi:10.1016/0016-7185(85)90029-6

Hollanders H., Tarantola S. and Loschky A. (2009) *Regional Innovation Scoreboard (RIS) 2009.*

Johansson B. (1993) Economic evolution and urban infrastructure dynamics, in Andersson Å. E., Batten D. F. Kobayashi K. and Yoshikawa K. (Eds) *The Cosmo-Creative Society*, pp. 151–175. Springer, Berlin.

Johansson B., Karlsson C. and Stough R. (2001) Theories of endogenous regional growth – lessons for regional policies, in Johansson B., Karlsson C. and Stough R. (Eds) *Theories of Endogenous Regional Growth – Lessons for Regional Policies*, pp. 406–414. Springer, Berlin.

Kaldor N. (1970) The case for regional policies, *Scottish Journal of Political Economy* **17**, 337–348. doi:10.1111/j.1467-9485.1970.tb00712.x

Kogler D. F. (2015) Editorial: Evolutionary economic geography – theoretical and empirical progress, *Regional Studies* **49**, 705–711. doi:10.1080/00343404.2015.1033178

Leydesdorff L. and Etkowitz H. (1998) The triple-helix as a model for innovation studies, *Science and Public Policy* **3**, 195–203.

Lundvall B.-A. (Ed.) (1992) *National Systems of Innovation.* Pinter, London.

Moen J. (2005) Is mobility of technical personnel a source of R&D spillovers?, *Journal of Labor Economics* **23**, 81–114. doi:10.1086/425434

Myrdal G. (1957) *Economic Theory and Underdeveloped Regions.* Duckworth, London.

Nelson R. R. (Ed.) (1993) *National Innovation Systems: A Comparative Analysis.* Oxford University Press, Oxford.

Nelson R. R. and Rosenberg N. (1998) Science, technological advance and economic growth, in Chandler A. D. Jr, Hagstrom P. and Solvell O. (Eds) *The Dynamic Firm: The Role of Technology, Strategy, Organisations and Regions.* Oxford University Press, Oxford.

Nelson R. R. and Winter S. G. (1982) *An Evolutionary Theory of Economic Change.* Harvard University Press, Cambridge, MA.

Organisation for Economic Co-operation and Development (OECD) (2011) *Regions and Innovation Policy.* OECD Reviews of Regional Innovation. OECD, Paris.

Polanyi M. (1969) *Knowing and Being*, ed. and intro. by Grene M. University of Chicago Press, Chicago, IL.

Richardson H. W. (1978) *Regional Economics.* University of Illinois Press, Urbana, IL.

Rodríguez-Pose A. (1999) Innovation prone and innovation averse societies: economic performance in Europe, *Growth and Change* **30**, 75–105. doi:10.1111/0017-4815.00105

Rodríguez-Pose A. (2001) Is R&D investment in lagging areas of Europe worthwhile? Theory and empirical evidence, *Papers in Regional Science* **80**, 275–295. doi:10.1007/PL00013631

Rosenberg N. (1982) *Inside the Black Box: Technology and Economics.* Cambridge University Press, Cambridge.

Sterlacchini A. (2008) R&D, higher education and regional growth: uneven linkages among European regions, *Research Policy* **37**, 1096–1107. doi:10.1016/j.respol.2008.04.009

Usai S. (2011) The geography of inventive activity in OECD regions, *Regional Studies* **45**, 711–731. doi:10.1080/00343401003792492

Varga A. (Ed.) (2009) *Universities, Knowledge Transfer and Regional Development: Geography, Entrepreneurship and Policy.* Edward Elgar, Cheltenham.

Wolstenholme E. F. (1990) *System Enquiry.* Wiley, Chichester.

Index

Page numbers in italics refer to figures. Page numbers in bold refer to tables.